The Institute of Biology's
Studies in Biology no. 98

Plant Physiological Ecology

John R. Etherington
Ph.D., D.I.C.
Senior Lecturer in Botany,
University College of South Wales,
Cardiff

Edward Arnold

First published 1978
by Edward Arnold (Publishers) Limited
41 Bedford Square, London, WC1B 3DP

Boards edition ISBN: 0 7131 2689 2
Paper edition ISBN: 0 7131 2690 6

Printed and bound in Great Britain at
The Camelot Press Ltd, Southampton

General Preface to the Series

It is no longer possible for one textbook to cover the whole field of Biology and to remain sufficiently up to date. At the same time teachers and students at school, college or university need to keep abreast of recent trends and know where the most significant developments are taking place.

To meet the need for this progressive approach the Institute of Biology has for some years sponsored this series of booklets dealing with subjects specially selected by a panel of editors. The enthusiastic acceptance of the series by teachers and students at school, college and university shows the usefulness of the books in providing a clear and up-to-date coverage of topics, particularly in areas of research and changing views.

Among features of the series are the attention given to methods, the inclusion of a selected list of books for further reading and, wherever possible, suggestions for practical work.

Readers' comments will be welcomed by the author or the Education Officer of the Institute.

1978

The Institute of Biology,
41 Queen's Gate,
London, SW7 5HU

Preface

Ecology may be defined as the study of the complex interrelationship between organisms, their environment and each other. This book is about the province of physiological ecology: the interpretation of plant behaviour and distribution in terms of physiological responses to the environment. Subject to competition, pests, pathogens and environmental stresses, these responses are often very different from those of plants raised under optimal environmental conditions, usually because competitive exclusion limits species to habitats which are not entirely favourable. The critical study of plant behaviour under these conditions demands a wide range of techniques.

The scope of the subject is such that its relevant literature is scattered through a host of specialist publications and is not comprehensively discussed in any elementary texts. The aim of this book is to fill this gap.

Cardiff, 1978

J. R. E

Contents

1 Introduction

1.1 What is ecology?

From very early times man noticed that different sorts of plants grow in different places: rushes and sedges in the wet hollows, bracken on the hillsides, grass and heather on the windswept summits. The common names of many wild plants were given in recognition of this fact: wood anemone; marsh marigold; field pansy; sea holly; mountain ash and meadowsweet are but a few. Every farmer and gardener likewise understands that his plants have different tolerances or preferences: for acid or alkaline soil, for sand or heavy clay and for wet or dry, sunlit or shaded conditions.

The cultivation of soil is a constant battle, at first against the teeming seed-life of newly broken ground, usually weed species which have lain dormant since some previous disturbance. The newly ploughed field becomes yellow with the flowers of charlock *(Sinapis arvensis)* and the interlacing stolons of creeping buttercup *(Ranunculus repens)* reclaim the laboriously won garden. Later, opportunist invaders with creeping rhizomes or budding roots gain a foothold. Notorious examples are the rhizomatous couch grass *(Agropyron repens)* and the field thistle *(Cirsium arvense)*.

Wondering how and why these things happen men must have nurtured ecological thoughts in their first efforts to gather food, to hunt animals and to domesticate plants but it was not until the late nineteenth century that such ideas gave birth to the infant science of *Ecology*. *Plant Ecology* is a more limited field: its definition is obvious but the reason for its existence is less easy to justify. Unfortunately there are few who are academically equipped to cope with both animal and plant biology and consequently the separation persists. Despite this, ecology has been the field for more interdisciplinary study than most other branches of science.

The subject of this book, *Plant Physiological Ecology*, seeks to extend the scope of ecology beyond the verbal and numerical description of plant distribution in the field. Such studies, at best, can only establish correlations between the distribution of species or groups of species and a multiplex of environmental factors. In a sense, the field ecologist is looking at the end results of gargantuan 'natural experiments' and can only study the individuals left by the survival of the fittest. He may speculate upon the reasons for the presence or absence of particular species but the experimental techniques of physiological ecology must be introduced if proof of causality is sought.

The physiological study of individual plant function and its relationship to environment, the subject matter of classical plant physiology, is so closely related to some aspects of plant ecology that it is not possible to define the boundary between them. Most of the techniques of physiological ecology are similar to those of physiology though the extension of physiological method to field conditions may require a modified approach to cope with lack of environmental control. The collection of environmental data in the field has, however, demanded a new technology of micrometeorological instrumentation and the development of measuring and monitoring techniques for soil chemical and physical conditions.

It is difficult to establish who first used the term physiological ecology but, as early as 1917, TANSLEY wrote of an experiment with two *Galium* spp. 'No attempt was made to analyse the cause of the effect of calcareous soil on *Galium saxatile*. . . . Such an attempt would form the subject of an interesting investigation in physiological ecology.' It is interesting to note that CLEMENTS, twelve years before this (1905) had deplored the fact that ecology and physiology were not synonymous terms. He proposed that ecologists should adopt the more precise techniques of plant physiology and that plant physiologists '. . . should recognize more fully that function is the middleman between habitat and plant'.

2 Energy Sources

2.1 Energy and life

Living organisms exploit the energy sources of the biosphere to support and reproduce themselves, creating local accumulations of organized matter and energy and it may be that one interpretation of natural selection is that it maximizes useful energy flow (H. T. ODUM, 1971).

Energy may manifest itself as the kinetic energy of a moving body or in the potential energy of a mechanical or chemical store. It may also be propagated through space as electromagnetic radiation. All forms may be interconverted and it is by such conversions that organisms obtain the energy which is necessary for their survival, reproduction and evolution.

The use of energy by organisms and its flow through ecosystems is governed by the first two Laws of Thermodynamics which state that:

(i) the sum of all of the energy in an isolated system is constant; and
(ii) all systems tend to approach a state of maximum probability (maximum disorder): spontaneous changes are those which proceed in this direction.

The first law implies that energy can neither be created nor destroyed and that energy input to the ecosystem must be equalled by storage plus output. The energy flow or energy balance approach to ecosystem analysis is based upon this assumption. The second law may broadly be interpreted to suggest that complex functional systems and organisms will spontaneously deteriorate and that energy will be required to maintain their order. Both organisms and ecosystems thus have a basal metabolic requirement to prevent deterioration with time.

Organisms utilize energy derived from two sources. The first is the chemical potential energy which may be drawn from inorganic or organic materials and the second is the radiant energy of the visible part of the electromagnetic spectrum. This is light energy which is harnessed to ecosystem power through photosynthetic carbon assimilation. The range of energy trapping processes is summarized in Fig. 2–1. At the present day the bulk of global ecosystem energy flow is initiated by photolithotrophs and most of the photosynthetically trapped energy is handed on through a food web of chemoorganotrophic consumers and decomposers. The weight of carbon fixed annually has been estimated at about 0.7×10^{11} metric tons: without respiratory return this would exhaust the atmospheric and oceanic dissolved carbon dioxide supply within about 200 years!

The contribution which photosynthesis makes to the biogeochemical

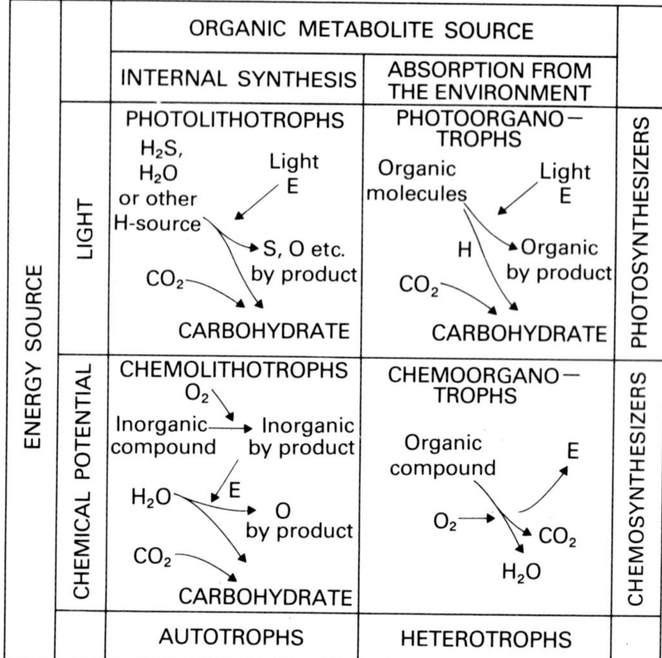

Fig. 2–1 Exploitation of environmental energy sources by organisms.

cycle is also witnessed by our oxygen-rich atmosphere which is believed to be entirely biogenic. The appearance of the photosynthetic biochemical pathway, some three thousand million years ago, provided the first extensive source of atmospheric oxygen which, from that time, increased steadily to the present level of about 21% v/v. Carbon dioxide was concurrently reduced in concentration until today, in equilibrium with global photosynthesis, respiration and the buffer of solubility in the oceans, it is only 300–350 volumes per million (vpm). (See section 4.4.)

The sun, with an apparent surface temperature of 6000 K, radiates an almost unchanging stream of energy into the space around it. So great is this radiation that earth receives it with a flux density of 1360 watts per square metre (W m^{-2}), the solar constant, at the sunward limit of the upper atmosphere. This corresponds roughly to the energy output of one small electric fire per square yard!

2.2 Radiant energy and energy balance

All objects at temperatures above absolute zero emit energy as electromagnetic radiation, the wavelength of which is inversely related to absolute temperature.

The amount of energy emitted is a function of the fourth power of absolute temperature as defined by Stefan's law:

$$B = \sigma T^4$$

where B = radiant flux density (W m^{-2}); T = absolute temperature (K); σ = Stefan-Boltzmann constant (5.57×10^{-8} W m^{-2} K^{-4}).

Very hot bodies thus radiate enormous amounts of energy compared with relatively cool objects. The spectral distribution of the radiation is defined by Wein's law:

$$\lambda_{max} = \frac{2897}{T}$$

where λ_{max} = wavelength at which the energy content per unit wavelength is greatest (μm).

A very hot object like the sun (6000 K) radiates about half its energy in the short-wave visible part of the spectrum (Fig. 2–2) but, because the spectral distribution of wavelength is skewed about the mode, the other half forms a long tail in the near infra-red. Figure 2–2 shows the great temperature sensitivity of the energy spectrum. An object at 2000 K

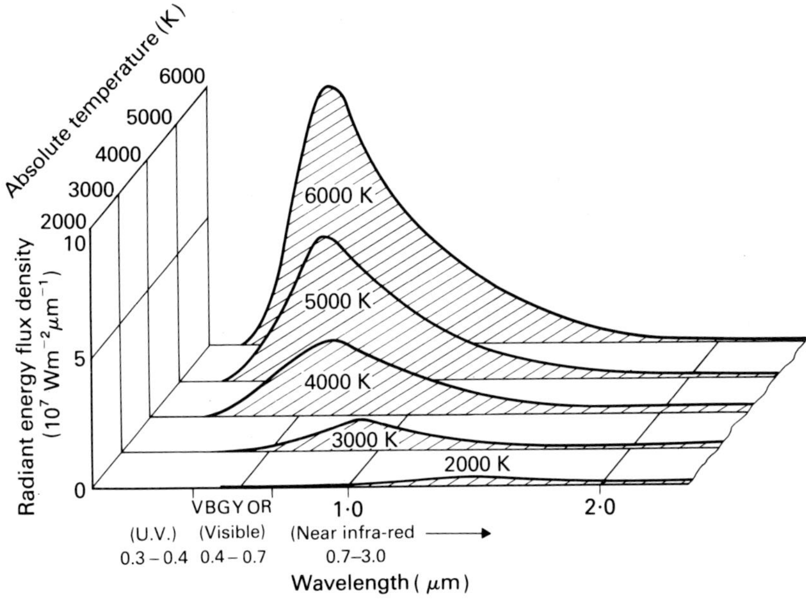

Fig. 2–2 The effect of surface temperature on spectral composition and energy content of electromagnetic radiation from a 'black-body' radiator. The highest temperature shown corresponds to the surface of the sun which consequently radiates with a spectral peak in the yellow-green part of the visible.

though visually white-hot radiates almost all of its energy in the near infra-red (0.7–3.0 μm). The earth with an apparent extra-terrestrial temperature of between 250 and 300 K emits entirely in the far infra-red (3–100 μm).

A body which emits radiation is equally capable of absorbing it. A *black body* is one which has an *emittance* and an *absorbance* of unity. If the body is less than perfectly 'black', Stefan's law must be modified to $B = E\sigma T^4$ where the emittance, E, is a constant of value 0–1.0. In a closed system all surfaces will be emitting and absorbing according to their respective temperatures and emittance – absorbance characteristics. The net amount and direction of energy exchange will be a function of these characteristics. More detailed reviews may be found in GATES (1962) and MONTEITH (1973). The earth's surface receives radiant energy from the sun in the spectral waveband of 0.3–3.0 μm at a maximum flux density which rarely exceeds 75% of the solar constant (1360 W m^{-2}) consequent on absorption and reflection by atmospheric gases, water droplets and dust. The solar input is absorbed by the atmosphere or by the earth's surface and manifests itself as heat or as latent heat storage. Less than 1% is diverted into chemical potential energy by photosynthetic storage.

Receiving a continuous energy from the sun, why does the earth not steadily rise in temperature? The answer is found in Stefan's law which shows that there will be an equilibrium temperature at which the earth will be radiating as much energy to space as it is receiving from the sun. This is long-wave (thermal) infra-red in the waveband 3.0–100 μm. Most of the long-wave loss is from the atmosphere as the long-wave radiation from the terrestrial surface is blanketed by the atmosphere (greenhouse effect). The mean global temperature is about 288 K representing an input : output balance of 383 W m^{-2}. Figure 2–3 shows the main pathways of the global energy balance.

During the daytime, plants, animals and physical surfaces are subject to an input of short-wave (0.3–3.0 μm) solar radiation and long-wave (3–100 μm) input from the atmosphere and surroundings. Consider a horizontal surface which is receiving sunlight at a flux density of 1000 W m^{-2} and losing heat only by long-wave reradiation. Stefan's law shows, assuming the surface to be a perfect radiation absorber, that its equilibrium temperature would be more than 90°C. This is well illustrated by the intense heat of a black car parked in the sun. Even with convective cooling these temperatures are well above lethal levels for most organisms.

Plants, unlike animals, cannot move to escape solar heating but they have a number of morphological and physiological attributes which prevent them from, quite literally, being cooked. Selection has favoured the thin leaf in most plants, providing a short carbon dioxide diffusion path to the mesophyll cells and maximum exposure of chlorophyll to incoming light. Such free diffusive gas exchange also makes the leaf a very efficient water evaporator, hence the need for the stomatal closure

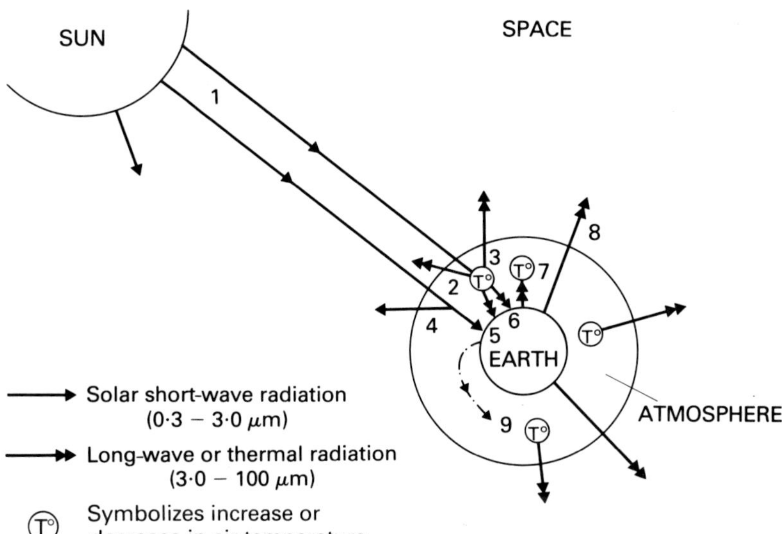

Fig. 2–3 The radiant energy balance of earth. Solar short-wave radiation (1) enters the upper atmosphere (2). It may be absorbed (3) causing atmospheric heating, or reflected (4) and lost to space. The remainder is transmitted to the terrestrial surface (5). Long-wave radiation from the heated atmosphere also reaches the surface (6). The absorption of this short and long-wave radiation leads to surface heating, evaporation of water and melting of ice. Long-wave radiation from the surface is absorbed by the atmosphere (7) or escapes to the space sink (8). The majority of the long-wave loss to space is, however, from the atmosphere (3). Direct radiant heating of the atmosphere and conduction and convection at the terrestrial surface provides the energy which drives earth's atmospheric circulation ensuring continuity of the water cycle and other gaseous elemental cycles (9).

mechanism. Transpiration from a square metre of vegetation may easily reach 0.5 kg h^{-1} in dry, sunny weather. This is a latent heat flux of almost 350 W m^{-2} and would be nearly half the radiant heat loading in bright sunlight.

The thin leaf has a very low heat capacity per unit area and, consequently, is tightly coupled to air temperature through convective or forced convective air cooling. Other factors contributing to the reduction of radiant heat loading are the development of highly reflective epidermal hair coatings on some leaves and the low absorbance of most leaves in the near infra-red (0.7–1.25 μm).

Intense sunlight combined with still air and high humidity may, however, cause considerable leaf heating. Excess temperatures of 10–20°C are common and, with air temperatures above 30°C, can result in physiological damage. The mid-day photosynthetic slump, often observed on hot sunny days, has been attributed to overheating which may either inflict biochemical injury or stimulate respiratory carbon dioxide production causing the guard cells of the stomata to close. Midday stomatal closure has, however, also been attributed to transient leaf water deficit.

2.3 Ecosystem water loss

At night, upward loss of long-wave radiation to the sky exceeds downward income from the atmosphere, the net radiation flux is negative and the ecosystem tends to cool. Except for the removal of heat from air adjacent to the leaves no energy is available for transpiration but, after sunrise, the input of solar short-wave energy soon brings net radiation flux to a positive value and provides the energy for ecosystem water loss.

Because it is quite difficult to monitor transpiration from whole ecosystems, various workers have attempted to calculate potential water loss from measurements of ecosystem energy balance. These calculations of *potential transpiration*, based on the latent heat of vaporization energy requirement, are sufficiently accurate for irrigation control and hydrological catchment yield prediction. Full discussion may be found in PENMAN (1963).

Vegetation type seems to make little difference to ecosystem water loss once the canopy is closed and if plants of similar size are compared. There is, however, some evidence that forests may transpire 10–20% more water than grasslands when water supply is unlimited. This has some significance in catchment hydrology (SWANK and DOUGLAS, 1974) and the consequent water-logging has also been suspected of initiating soil deterioration and peat formation after tree clearance in wet areas (DIMBLEBY, 1967). The average water loss from a closed vegetation cover is about 0.8 of the free water surface value but this may be exceeded in small vegetation stands if warm, dry air is carried in from adjacent unvegetated areas ('oasis effect').

2.4 Photosynthetic production and energy flow

Net primary production may be defined as total dry weight production per unit area per unit time. It represents gross photosynthesis minus internal respiratory loss. Nutrient element uptake may be neglected as being a small proportion of primary production. Globally, primary production is strongly correlated with climate through direct and indirect effects of radiant input and is the dominant energy source for all other ecological processes. The highest yielding tropical crops, such as sugar cane, may produce almost 7 kg m^{-2} $year^{-1}$ of dry weight, an energy fixation of more than 100 MJ m^{-2} $year^{-1}$ and efficiency of about 4% if the annual radiant energy input is assumed to be 3000 MJ m^{-2} in the photosynthetically usable waveband.

Even this efficiency is unlikely to be achieved in wild ecosystems. E. P. ODUM (1971) cites figures ranging from 0.5% for fertile ecosystems down to 0.05% for the open ocean and for semi-arid vegetations. He suggests 4% as a maximum and about 0.1% as a whole-biosphere annual efficiency. These estimates are based on annual *total* radiant input. More recently a compilation of International Biological Programme figures has been published (COOPER, 1975; LEITH and WHITTAKER, 1975). Net primary production figures range from as little as 0.01 kg m^{-2} for tundra and semi-arid areas up to 3 or more kg m^{-2} $year^{-1}$ in tropical rain forest. Temperate grasslands are intermediate and similar to summer deciduous forest (about 1 kg m^{-2} $year^{-1}$). This more recent work suggests a biosphere annual efficiency of 0.2% of the *photosynthetically available* radiation and a maximum short-term efficiency of 3–4% for crops during the growing season. These very low solar energy conversions are the basis of the current preoccupation with increasing agricultural production efficiency.

The poor photosynthetic efficiency of the vegetation canopy is compounded by a series of losses which occur with each energy transfer in the ecosystem. The consumption of primary producers by herbivores, their subsequent predation by carnivores and the processing of organic molecules by microorganisms all involve energy transfer and its loss to the environment as heat. The transfer loss represents perhaps 90% at each step so that energy flow through the final carnivore levels of an ecosystem may be no more than 0.1% of the primary production and usually less than 0.0005% of the radiant input to the system.

It is perhaps unwise to use the term 'loss' to describe the reduction of energy flow at each transfer. The surplus energy appears as environmental heat, generated in respiratory metabolism and is the thermodynamic consequence of the extraction of useful work from metabolic energy stores. Thus the 'loss' is in fact the payment for the creation and maintenance of ordered structures at both organism and ecosystem level.

The concept of energy flow and efficiency has been discussed extensively elsewhere in this series (PHILLIPSON, 1966) and requires but brief attention here. Energy-flow studies are usually considered to have been initiated by LINDEMAN (1942) who first drew attention to the dynamics of ecological food chains. By defining the various *trophic levels* of these chains and by utilizing the physical concepts of energy content and transfer, this work laid the foundation of most subsequent attempts at ecosystem analysis.

Pioneered by ODUM (1957) and his co-workers, the construction of energy-flow models began the development of ecosystem analysis which is now one of the fast growth sectors of modern ecology. Concepts of efficiency may be incorporated in such models, for example the flow-diagram of Fig. 2–4 shows the quantitative relationships of biomass and energy transfer in a mature forest ecosystem. A great deal of gross photosynthetic production is dispersed at the first trophic level in respiratory maintenance and the remainder supports an equilibrium population of herbivores, carnivores and decomposer organisms, leaving no surplus for storage. In this case a large primary producer biomass is capable of supporting only a relatively limited consumer biomass to which is passed but a small proportion of the ecosystem energy flow.

By contrast, some ecosystems have a small primary producer biomass which is constantly cropped and little energy is devoted to maintenance respiration despite the high energy throughout. This is typical of many aquatic systems in which the plants need little mechanical support tissue and molluscan browsing or filter feeding constantly rejuvenates the plant cover.

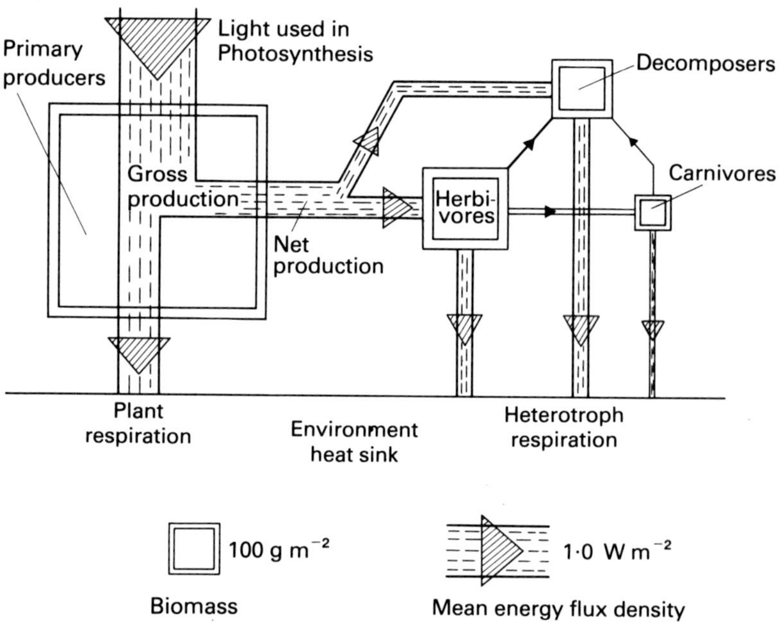

Fig. 2–4 Energy-flow diagram for a climax forest ecosystem. An equilibrium has been attained in which the whole of the gross primary production is used up by plant and heterotroph respiration. The width of the flow-paths is intended to be proportional to energy-flux density and the areas of the 'boxes' to biomass. The energy-flux density is based on the annual mean assuming a gross photosynthetic production of about 1% of the solar input. In the temperate zone the solar short-wave input, averaged over a year, is approximately 8.8 MJ m^{-2} day^{-1}, equivalent to a flux density of 102 W m^{-2}. In an immature system further flow-paths would extend from all trophic levels to a storage as either increasing biomass or organic debris.

Young ecosystems usually have a net primary production which is in excess of consumption and are consequently increasing in biomass. It is suggested that as succession proceeds towards the climax condition, biomass approaches a maximal value. This successional change is probably also accompanied by maximization of the rate and capacity of the nutrient cycle (section 4.1). Extensive discussion of these concepts may be found in E. P. ODUM (1971).

3 Material Sources: Atmosphere and Soil

3.1 The atmosphere

The outer planets of the solar system are known to have atmospheres containing water, methane, carbon dioxide and ammonia; chemically reducing systems in which free oxygen cannot exist. The primary atmosphere of earth is believed to have been similar and formed the carbon-rich environment in which life originated. This first evolutionary step permitted the development of microorganisms which were dependent on local chemical energy sources (see Fig. 2–1). Without the biochemical evolution of the photosynthetic process, the biosphere might still be dominated by a handful of chemotrophic microbial species with a very small annual carbon budget.

The surface of our planet and its atmosphere form a system of limited size in relation to the impact of the organisms which it carries. Our oxygen-rich, carbon-poor atmosphere has been generated by photosynthesis and the proponents of the 'spaceship earth' concept suggest that if the human population is not to degrade its own environment, the atmosphere must be treated as a biologically regenerated system upon which our agriculture and technology now have a significant effect. With the aid of scientific and technical advance, fossil fuel has amplified our ability to interfere with a system which has taken perhaps three to three and a half thousand million years to evolve to its present condition.

3.2 Climate and microclimate

The daily revolution of the earth creates the short-term day-night cycle within the seasonal pattern, and variation of cloud cover superimposes both random changes and longer-term cyclic changes. The variations of radiant input, temperature and climatic wetness have a strong influence on the geographical distribution of vegetation, soils and primary productivity (Fig. 3–1).

The terrestrial plant is a biological bridge between two very contrasting parts of the environment. Materials move in soil by flow of soil solution as thin capillary films between soil particles and by molecular diffusion in the gaseous or liquid phase. By contrast with these slow processes, molecular movement in air is fast, as a result of wind and local convection currents. The soil : air interface, without vegetation cover, is consequently a site of extreme environmental discontinuities. For example the *diurnal*

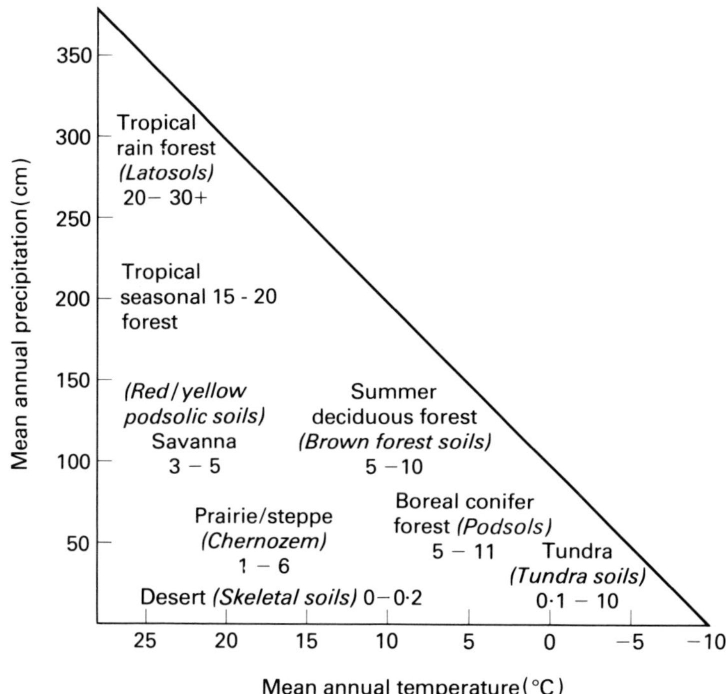

Fig. 3–1 The relationship of vegetation types, soils and annual net primary production (tonnes hectare^{-1}) to climatic temperature and wetness. Mean production figures based on estimates given in COOPER (1975).

range of soil surface temperatures may often exceed the *annual* range of mean air temperature.

The physical presence of vegetation smooths the sharp gradients of temperature and air-speed and its physiological activities create diurnally changing vertical gradients of water vapour and carbon dioxide concentration. Figure 3–2 shows some generalized profiles of these microclimatic components and their daily range of variation. Measurements of carbon dioxide and water vapour concentration gradients above and within the canopy may be used to calculate whole ecosystem photosynthesis, respiration and water loss (MONTEITH, 1973). The ecological significance of micrometeorological measurements lies not so much in its potential use for measuring fluxes but more in its ability to specify the detailed environmental characteristics to which individual plants or plant organs are exposed within the canopy.

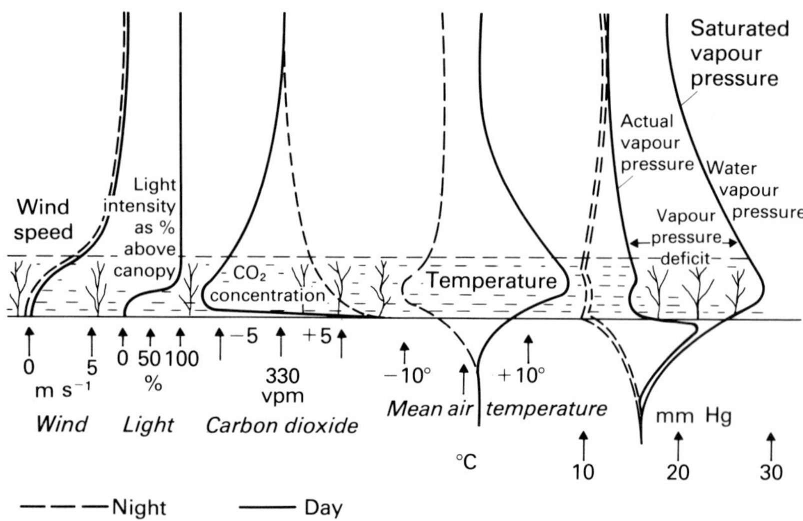

Fig. 3–2 Micrometeorological profiles through a vegetation canopy. Wind speed is shown for light breeze conditions but the maximum speed can occasionally rise to more than 25 m s^{-1} (whole gale). Under these circumstances the speed at ground level would be considerably above zero. Light intensity is plotted as a percentage of the incident intensity at the top of the canopy. Some herbaceous communities may reduce the intensity at ground level to as little as 1–5%. The carbon dioxide profiles are plotted for low wind-speed conditions and show the daytime reduction of CO_2 concentration in the canopy caused by photosynthetic uptake in high sunlight conditions. The night-time increase is caused by respiration of the canopy and above ground organisms. The high concentration at ground level is a consequence of soil animal, microorganism and root respiration. The diurnally changing temperature profile reflects the absorption of short-wave solar energy during the day and the upward loss of long-wave infra-red radiation at night. The gradients in and above the canopy are a consequence of convective and conductive heat exchange and the exchange of latent heat during transpiration or nocturnal dew formation. The saturation vapour pressure curves for water are closely related to those of temperature as the capacity of air to contain water vapour is very strongly temperature sensitive. At night the cooling air often reaches saturation and dew may be deposited on plant and soil surfaces. During the day, by contrast, the warmer air is often far from saturation and provides the sink for transpired water vapour. Below the soil surface the temperature curves approach a mean temperature which reaches a maximum in later summer and a minimum in late winter. The saturation vapour pressure curves below ground follow this temperature pattern but the daytime actual vapour pressure curve reaches saturation because air movement is restricted.

3.3 The soil

Seasonal and diurnal changes of climate and microclimate will cause a bare rock surface to be broken down by physical and chemical weathering even in the absence of life. When it is invaded by autotrophic microorganisms such as blue-green and green algae or by lichens with their photosynthetic algal symbionts, the process is greatly accelerated. For example, various metabolites such as organic acids increase the rate of chemical weathering while the appearance of heterotrophic organisms supported by algal photosynthesate increases the diversity of organic materials contributing to the processes which form soil. Accumulation of organic matter will modify surface physical conditions, in particular increasing water-retaining capacity, assisting the establishment of more ecologically demanding higher plants.

When the input of fresh organic matter reaches equilibrium with the breakdown process the soil may be considered to have reached maturity. This will occupy a long period of time and may be accompanied by vegetational succession. PAUL *et al.* (1964), for example, showed that humic acids from a grassland soil had a mean radio-carbon age of about 1000 years. This gives some impression of the stability of the soil organic matter molecules, the time-scale of soil formation and highlights the need for soil conservative management in agriculture. Reclamation of derelict sites and restoration of deteriorated soils also pose the problem of recreating a mature soil in an abnormally short time.

In an ecological context soil may be defined as the material in which plants root and from which they draw their water supplies and essential mineral nutrient elements. A normal soil is a three-phase, solid : liquid : gas, system of variable constitution (Fig. 3–3). Over short periods of time the inorganic and organic solid component does not vary but the soil atmosphere and soil solution alter inversely in volume as water enters or leaves the soil. The composition of the soil atmosphere and soil solution may also differ according to environmental circumstances.

The mineral particles of the soil may be classified into size-classes which though arbitrary, do separate the finest *clay* fraction (<0.002 mm equivalant diameter) from the larger *sand* and *silt* components at a point on the particle-size continuum which coincides with a marked change of physical-chemical characteristics. The larger particles are relatively inert quartz and unweathered mineral grains contrasting with the layer-lattice crystals of alumino-silicate clays (Fig. 3–4) of which the chemical activity is partially due to their enormous surface : volume ratio and partially to their surface negative charge caused by isomorphous replacement of aluminium and silicon ions in the crystal lattice by ion species of lower valency (Fig. 3–4).

The surface negative charge confers cation exchanging properties upon

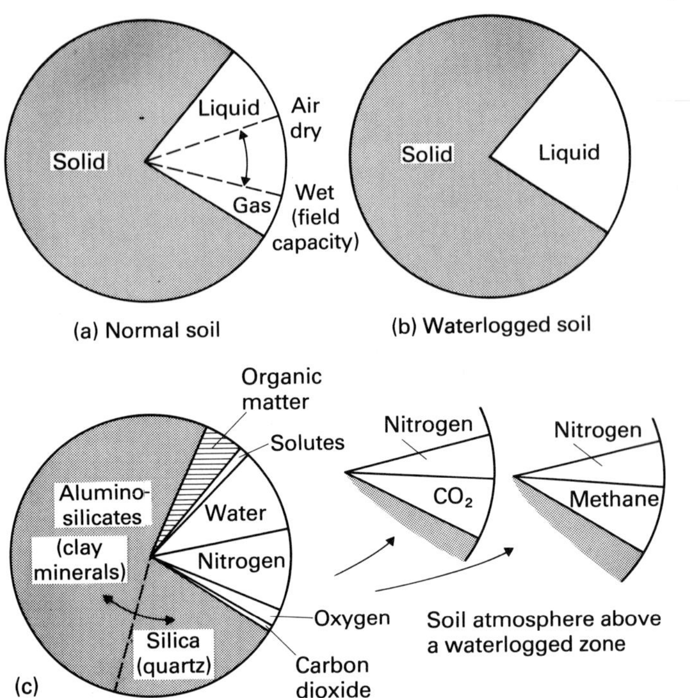

Fig. 3–3 The composition of soil. (**a**) shows the phase composition of a normal well-drained soil and (**b**) the replacement of the gaseous phase by water under flooded conditions. (**c**) gives a more critical breakdown of the constituents of a normal soil and the two sectors show the constitution of the soil atmosphere just above the water-table of a waterlogged soil. In one case the oxygen supply has been exhausted giving an atmosphere of nitrogen and carbon dioxide. In the second case anaerobic metabolism of organic matter has liberated a large amount of methane.

the clay minerals so that free metal cations and hydrogen ions in the soil solution become ionically bonded to the clay. Soil organic matter has strong cation exchanging properties and must also be considered in this context. In high rainfall areas *cation exchange capacity* provides protection against leaching loss of cations. Clay and organic matter replenish the supply of cationic nutrients after the soil solution is depleted by leaching or plant uptake thus buffering the soil against rapid changes of nutrient concentration or acidity. The clay minerals also provide the long-term reserve of plant nutrients for, geologically, it is from the alumino-silicates that most essential plant nutrients are derived.

The exchanging sites of the mineral and organic matter are described as the *cation exchange complex* and may be fully occupied by metal cations in

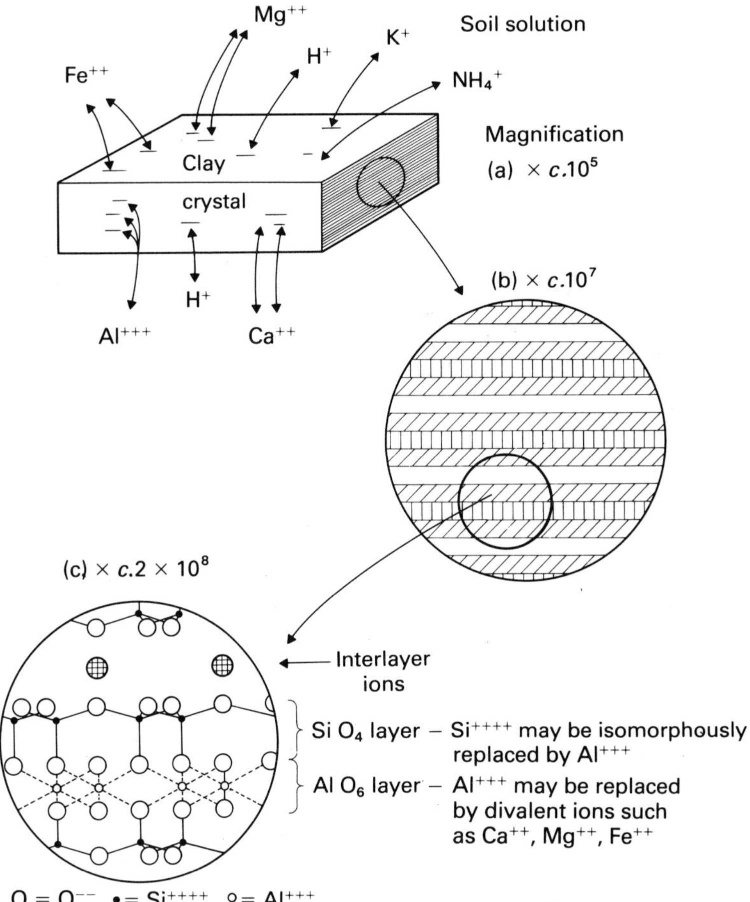

Fig. 3–4 (**a**) Diagram indicating the presence of surplus negative charge on the surface of clay crystals and its effect in exchanging cations from the surrounding soil solution; (**b**) shows the layer lattice structure of the clay and, at higher magnification; (**c**) illustrates the ionic structure of the layer-lattice and the way in which isomorphous replacement by elements of lower valency gives rise to the surplus negative charge.

which case the complex is said to be saturated. It was originally believed that sites unoccupied by metal cations carried *exchangeable hydrogen* ions which appear in solution if the soil is flushed with a neutral salt solution which displaces the hydrogen ions by substitution of metal cations. There is no way of differentiating between this process and one in which aluminium ions occupy the binding sites. On displacement into solution, aluminium ions would immediately precipitate as aluminium hydroxide with simultaneous release of hydrogen ions ($Al^{+++} + 3H_2O = Al(OH)_3 + 3H^+$) and it is now believed that acid soils contain aluminium rather than hydrogen ions on the exchange complex.

Irrespective of the mechanism of cation desaturation, the consequences of replacement of metal cations by either H^+ or Al^{+++} is an increase in soil acidity due to a higher concentration of H^+ in the soil solution. The pH of a soil slurry is conventionally measured with a calomel-glass electrode. The physical chemistry of a soil-water mixture is complex but, because soil pH varies widely (pH 3 to pH 10: see section 5.4.3) the measurement is ecologically meaningful (section 5.3.3 and Table 4). Temperate zone soils lie in the pH range from *c.* 3.0–8.4, a hydrogen ion concentration change of more than $\times 10^5$.

3·4 Soil organic matter

The soil contains a complex mixture of organic compounds. Some are derived from recently dead plants, animals and microorganisms but the complex soil organic macromolecules may be several hundred years old (section 3.3). Current biochemical opinion is that these molecules are composed of polyphenolic sub-units which are principally derived from lignin and plant flavonoids in the overlying vegetation.

The nature of the soil strongly influences the incorporation of organic matter. Neutral and alkaline soils of good nutrient status are biologically very active, containing large populations of earthworms, microarthropods, protozoa, bacteria and fungi. Plant litter, in these soils is rapidly converted to an amorphous *mull* humus which is colloidally associated with the mineral matrix throughout the full depth of the soil profile. By contrast, acid and nutrient deficient soils have few or no earthworms and a more limited bacterial population. The *mor* humus of these soils is not well mixed but remains as a fibrous layer at the surface. Mull is characteristic of forest and grassland *brown forest soils, chernozems* and limestone *rendzinas* while mor is associated with *podsols* and podsolic soils (BUNTING, 1965).

If waterlogging causes oxygen deficiency (section 5.3.3), organic matter may accumulate as *peat*. With a high precipitation : evaporation ratio (P/E) the peat extends over all horizontal surfaces and shallow slopes as a *blanket peat* which may reach a thickness of many metres. It is very nutrient deficient, having no capillary contact with ground water. By contrast, *valley peat*, which forms in hollows by accumulation of drainage water, is less nutrient deficient and, if the drainage water is alkaline and nutrient rich, may be replaced by *fen peat*. Under P/E conditions an alkaline fen peat may show surface acidification followed by the growth of *raised bog peat* with a similar flora to that of blanket peat. TANSLEY (1939) gives a good account of the formation of various bog types and more detail may be found in MOORE and BELLAMY (1973).

3.5 Soil formation

The combined influence of rainfall, evaporation and temperature governs many physical and chemical soil processes and also effects the course of succession and the climax vegetation. The physical and chemical constitution of the parent material has a strong influence and topography often has an overriding effect on soil type through the effects of waterlogging.

A high P/E regime maintains a downward movement of water through the soil and leaches soluble materials to the groundwater or to deeper horizons of the soil profile. The first materials to be mobilized are easily soluble ions and free calcium carbonate. Decalcification is often the first observable process in soil formation under these conditions.

If the soil is fairly coarse textured and freely drained, leaching not only removes nutrient elements faster than they are returned to the surface by biological cycling but also causes the mobilization of the usually insoluble iron, manganese and aluminium sesquioxides (R_2O_3). During the process of podsolization, the brown coatings of iron sesquioxide (Fe_2O_3) on mineral grains in the A-horizons are removed as organo-iron and redeposited in the deeper B-horizons. The bleaching of the surface mineral horizons and the red-brown colouration of the deeper soil with deposits of iron oxides are diagnostic characteristics of podsols.

Low P/E ratios promote capillary rise of water from deeper parts of the soil profile to satisfy evapo-transpiration demand. The rising water carries solutes such as calcium bicarbonate and sulphate which may enrich the surface layers. In some salt-desert soils there is continuous upward movement of salt-laden groundwater and the surface may become enormously enriched with sodium, potassium, magnesium and calcium chlorides, carbonates and bicarbonates.

The most obvious effect of temperature is on the accumulation of soil organic matter in well drained soils. Photosynthetic production is less temperature sensitive than the respiration of soil microorganisms. The consequence is a general gradient of increasing soil organic content from equator to sub-polar regions. The relationship is also related to the influence of temperature on evaporation. The high P/E ratios of the temperate and sub-arctic regions favour leaching, acidification and waterlogging-induced anaerobiosis with a consequent reduction of bacterial activity.

Because of these interacting factors soils may be very diverse both between and within profiles. Despite this diversity most soils provide the rooting medium for plants, supplying water and mineral nutrients in conditions of aeration and mechanical penetrability which allow the growth of at least some species.

A mixture of inorganic soil particles, without any organic content, have a dust-like character when dry. When wet, they are close-packed and of high bulk density, the smaller clay and silt particles filling the space between the sand grains. Such a soil would contain little water, be poorly aerated and resist the mechanical deformation caused by root penetration. In the field, the textural particles are, however, bound into *aggregates* both by the physico-chemical effects of soil organic molecules and by the physical binding effects of fungal hyphae and finer rootlets.

An aggregated soil is said to be *structured* and has large pores between aggregates which remain gas-filled unless the soil is entirely waterlogged. The pores also confer sufficient deformability to permit easy root penetration. Generally it is true to say that the activity of soil animals, particularly earthworms, is the major factor incorporating organic matter and producing structure under forest vegetation while the growth of roots with their large input of organic carbon is more important in grasslands.

3.6 Soil types

3.6.1 Soil profiles

The world-wide distribution of soil types shows a pattern which is strongly related to climate and vegetation. Locally, marked deviations may be a response to different rocks, topography or age of surface. Other local variations stem from unusual conditions such as flooding by sea-water.

Soils are described by reference to their *profiles*, the sequence of layers or *horizons* exposed in a section. Horizons are formed by biological and physico-chemical processes during the course of soil development and are usually identifiable visually but also differ from each other in chemical and physical properties.

Horizons differentiate in response to a number of biological and environmental effects. Soluble and finely particulate materials may be carried upward by the capillary rise of water under dry conditions, or downward by the infiltration of precipitation; plant nutrient elements are biologically returned to the surface as a constituent of plant and animal litter while soil animals, such as earthworms, mix soil from various depths, reversing the trend towards horizon differentation.

3.6.2 Podsols

In wet, temperate conditions, the dominant process is downward movement or *leaching* of soluble materials. If the soil becomes sufficiently leached the consequent acidification (section 3.3) excludes earthworms, litter begins to accumulate at the surface and horizon differentiation is accelerated.

The leaching process is enhanced by the liberation of polyphenolic organic compounds, from the litter. These mobilize the iron oxide coatings of the mineral grains which normally give brown or reddish soil colours. With strong leaching the mineral grains of the surface horizons may become totally bleached by downward loss of iron oxide (Fe_2O_3). Other sesquioxides (R_2O_3) such as aluminium and manganese may also be leached during this process of *podsolization*. The organo-iron and other complexes are microbially decomposed at some depth in the profile and produce a horizon of sesquioxide and organic matter enrichment. In some circumstances so much iron oxide is deposited that it forms a hard *iron-pan* (Fig. 3–5).

Extreme horizon differentiation of this kind is found in the soil type named a *podsol* (Fig. 3–6). The figure is labelled using a conventional notation in which the surface layers that are organic enriched and/or depleted of soluble and particulate materials by leaching, are called A-horizons. The deeper horizons enriched by inwashing of material from above are B-horizons and the horizon below this, essentially out of reach of major soil-forming processes, is the C-horizon. There is some difficulty in applying this notation to very wet soils and to various tropical soil types. Surface organic horizons are subscripted A_{00} (litter) and A_0 (strongly decayed organic layers). Podsols are usually temperate zone soils of nutrient-poor siliceous rocks under coniferous forest or Ericaceous heathland. Many heathland podsols are believed to be deterioration products of more fertile soils which were damaged by post-Bronze Age disafforestation which limited the depth and magnitude of cationic elemental cycling and promoted acidification. Podsols have a soil pH which is often below pH 4.0. The reduced mineral nutrient status coupled with the accumulation of a thick surface layer of *mor* humus (section 3.4) prevented forest regeneration, an effect which was reinforced by a greater fire risk and, in more recent years, by heavy grazing. Well differentiated podsols, such as that illustrated in Fig. 3–6a, are not particularly common in Britain but a range of podsolic soils are common in the northern and western uplands; the most common being various forms of *iron-pan podsol* (Figs. 3–5 and 3–6b) which form under wet conditions and are often associated with a superficial layer of peat.

3.6.3 Brown earths

The less siliceous rocks and glacial drifts of lowland Britain, with the exception of limestones, usually carry a mantle of brown earth type (Fig. 3–7). These soils show moderate to strong biological mixing, mainly by earthworms which prevents strong horizon development. Inherently nutrient-rich compared with the forest materials of podsols, these soils are maintained in a fertile condition both by soil mixing and by deep-seated elemental cycling through the forest cover which is their natural vegetation in Britain. They are usually near neutral with pH values between 6.0 and 7.0. The majority of these soils have now been taken for arable or grassland agriculture, the natural nutrient cycling being replaced by artificial fertilization. Addition of organic matter is often

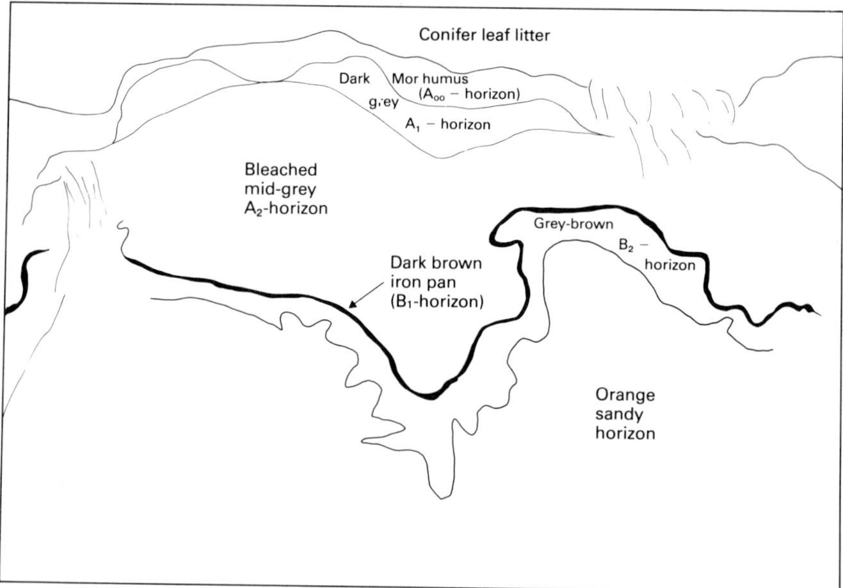

Fig. 3–5 An iron-pan podsol on Jurassic sandstone of the North Yorkshire moors. The natural vegetation is heathland dominated by ling (*Calluna vulgaris*) but this site has been afforested with various conifer species.

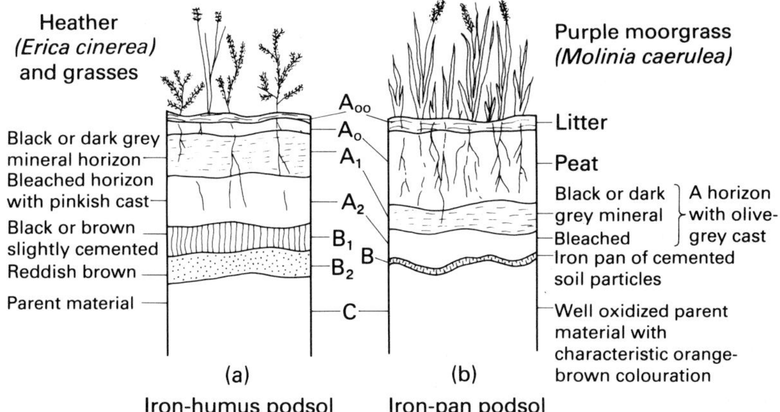

Fig. 3–6 A typical heathland podsol formed under well drained conditions (**a**) and an iron-pan podsol (**b**) formed under high rainfall : low evaporation circumstances typical of western and northern Britain on upland sites. The iron-pan acts as a barrier to water drainage and there is thus some gleying of the A horizons which accounts for their subdued olive-grey colours which contrasts with the warmer pinkish or brown tones of the iron-humus podsol A horizons.

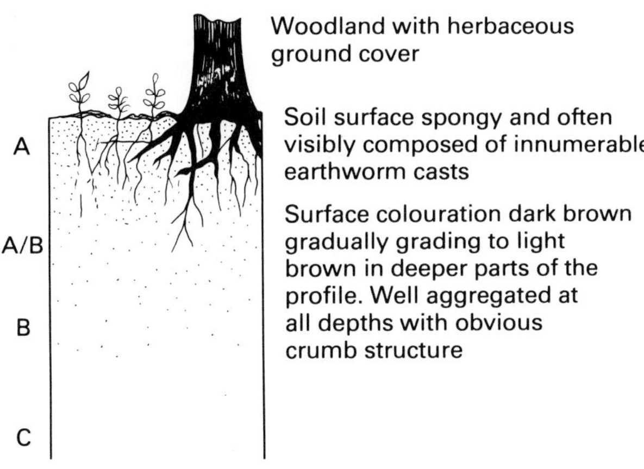

Fig. 3–7 A nutrient-rich brown earth or brown forest soil typically formed on lowland sites with non-siliceous, clay-rich parent materials. Characteristically dominated by oak (*Quercus robur*) or mixed deciduous woodland. Now almost entirely under arable or grassland agriculture. Some brown earths are formed on less favourable parent materials. These oligotrophic brown earths are more sandy and acid, less often under agricultural management and naturally would carry sessile oak (*Q. petraea*) and birch (*Betula pubescens* and *B. pendula*).

needed to maintain structure. The characteristic native woodlands of these soils were dominated by pedunculate oak (*Quercus robur*) on nutrient-rich lowland soils and sessile oak (*Q. petraea*) sometimes mixed with birch (*Betula pendula* and *B. pubescens*) on less fertile upland soils.

Earthworm activity, aided by that of many microscopic animals and a rich bacterial and fungal flora, ensures the rapid incorporation of fresh organic matter. The richest brown earth soils rarely have any surface litter layer except at the time of seasonal leaf-fall. The organic matter, intermixed with the soil mineral matrix forms a *mull* humus (section 3.4) which contributes to the well aggregated structure as well as providing cation exchange sites and forming a long-term reservoir of anionic nutrients such as nitrogen, phosphorus and sulphur.

Earthworms, in a nutrient-rich soil, may ingest between 50 and 100 t ha^{-1} y^{-1} of soil which is voided both in the pore spaces and at the surface. This constant turnover prevents a distinct differentiation of the A and B horizons of the brown earth, the organic content simply showing a gradual decrease with depth. The only sign of leaching is occasional enrichment of the B horizon with inwashed clay.

3.6.4 Gleys

Within the limits of normal parent materials and drainage, the brown earth and podsolic soil types are the most widespread in Britain but topogenic or climatic waterlogging may produce deviant wet soils called *gleys* in which oxygen deficiency limits decay to cause superficial accumulation of peaty organic matter and the deeper horizons show greyish or bluish-green colours mottled with the ochre and red-browns of oxidized iron components (Fig. 3–8). These gley layers (G-horizon) are formed by chemical and microbiological reduction of brown ferric iron compounds to grey or blue-green ferrous-organic complexes (section 5.3.3). These soils carry a range of plants tolerant to the various conditions caused by waterlogging (section 5.3.3).

Gleys may be considered to be wet variants of well-drained soil types of the same parent and consequently may have a wide range of acidity and nutrient status. They are usually, however, nearer to neutral (pH 7.0) than the related soil from which they are derived.

3.6.5 Rendzinas

Chalk and the harder limestones in Britain give rise to shallow soils in which an A-horizon rests directly on the C material (Fig. 3–9). They are alkaline (pH 7.5–8.4) because earthworm activity and solute movement maintains carbonate saturation to the surface. The soil organic matter is often very dark, staining the A-horizon black, dark brown or grey. In southern Britain the climax vegetation is beech (*Fagus sylvatica*) woodland but most typically it is now dominated by an anthropic, grazing-deflected

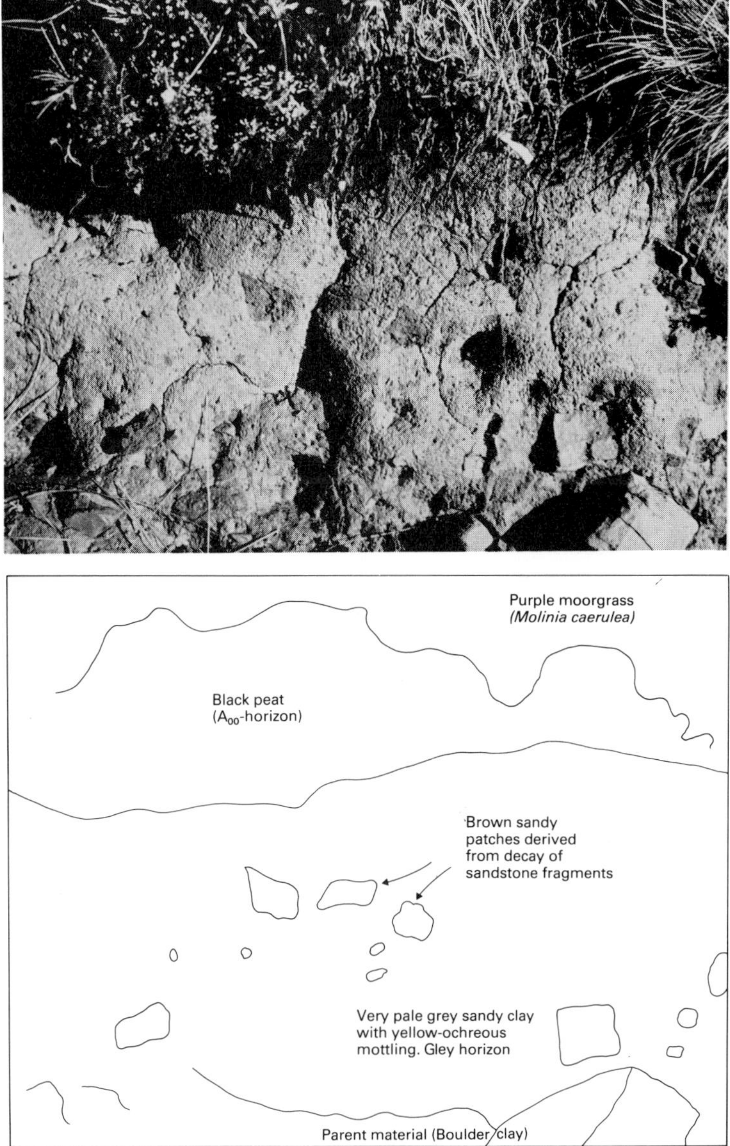

Fig. 3–8 A peaty-gley soil on boulder clay derived from Carboniferous shales and sandstones in South Wales. The characteristic mottling of the gley-horizon is not easily visible in a black and white photograph but heavy deposits of ferric compounds have been laid down in the sandy inclusions derived from decay of sandstone fragments.

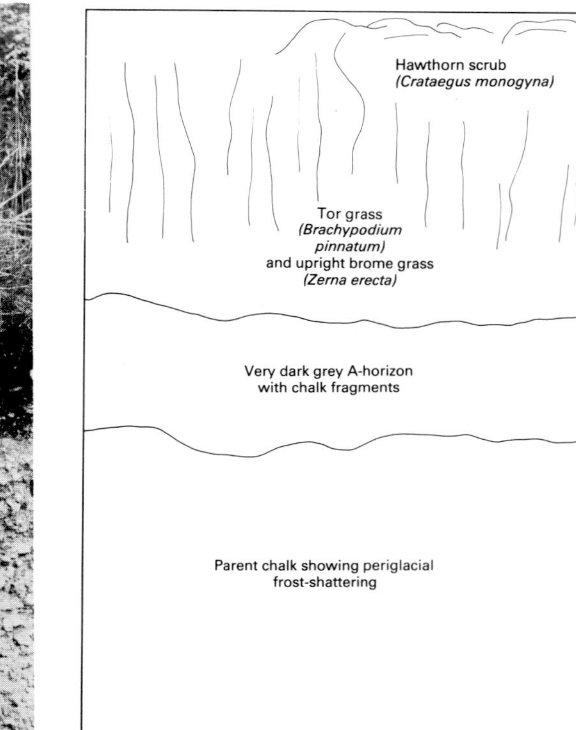

Fig. 3–9 A rendzina soil on Cretaceous chalk of the North Downs of Surrey. The very sharp boundary between the dark A-horizon and the parent chalk and the strong earthworm-mixing of chalk fragments is obvious.

climax of limestone grassland, the species-rich home of many of our calcicole plant species (sections 5.3.1, 5.3.4 and 6.2).

3.6.6 Other soils

Soils of tropical rain forest formed under conditions of high temperature and rainfall are very different from temperate soils in that silica (SiO_2) and most alumino-silicate clays are weathered to soluble compounds or colloidal silicic acid which is leached out of the soil causing residual enrichment with iron and aluminium sesquioxides (Fe_2O_3, Al_2O_3) and with kaolinitic clays. These *latosols* or *laterites* are the tropical red soils, such a problem in tropical agriculture because of their nutrient deficiency and tendency to set to a rock-like surface when exposed to sunlight and air (McNEIL, 1964).

Another frequent tropical soil type is the *tropical podsol* with an extremely deep bleached sandy A-horizon. The phenomenon of sesquioxide residual enrichment found in the latosols is also operative in the subtropics and in the weathering of calcium carbonate limestones under Mediterranean climate conditions to form the red *terra rossa* which are the counterparts of temperate zone rendzinas.

4 Material Cycles

4.1 Plants and elemental cycles

A climax ecosystem is generally considered to be one which has reached seral equilibrium of both species composition and standing biomass. The biomass of the climax system shows a slight seasonal cycle but has a constant annual mean, contrasting with the annually increasing biomass of the immature system. In this equilibrium situation the annual gross primary production is entirely expended in the respiration of plants, animals and microorganisms. If the ecosystem is not to deteriorate slowly, all of the material resources used in the synthesis of living tissue must ultimately be returned, after death, to the primary producers. A small loss to leaching or indefinite storage is tolerable if balanced by input in rainfall, blown dust or by release from freshly weathered geological material.

It should be clear that there is a close, reciprocal cause and effect relationship between energy flow through the ecosystem and cycling of materials within it (Fig. 4–1). As energy flow increases during the course of succession, more inorganic materials are needed to build living tissues but, at the same time, shortage of those same materials may act as a brake on both cycling and energy throughput. Shortage of inorganic raw materials is not the only reason for this sort of limitation: the rate of mobilization of organic stores of such elements as nitrogen, phosphorus and sulphur may also have a limiting effect.

In an ecological equilibrium, plant production may be limited by shortage of inorganic sources, low solubility of these sources or failure of microbial mineralization to keep pace with plant demand. If none of these factors limit photosynthetic production then the maximum rate of nutrient cycling will depend on some other environmental limitation to production such as availability of carbon dioxide, shortage of light or water, low temperature, pathological conditions or herbivore browsing.

4.2 The nature of the cycle

The biological cycles of chemical elements in the biosphere are often called *biogeochemical cycles* and concern the circulation of these essential elements which are absorbed in ionic form from soil solution, the structural elements – carbon, hydrogen and oxygen which are used in much larger quantities and, finally, a large number of elements which, though not essential, are absorbed from soil solution and concentrated in

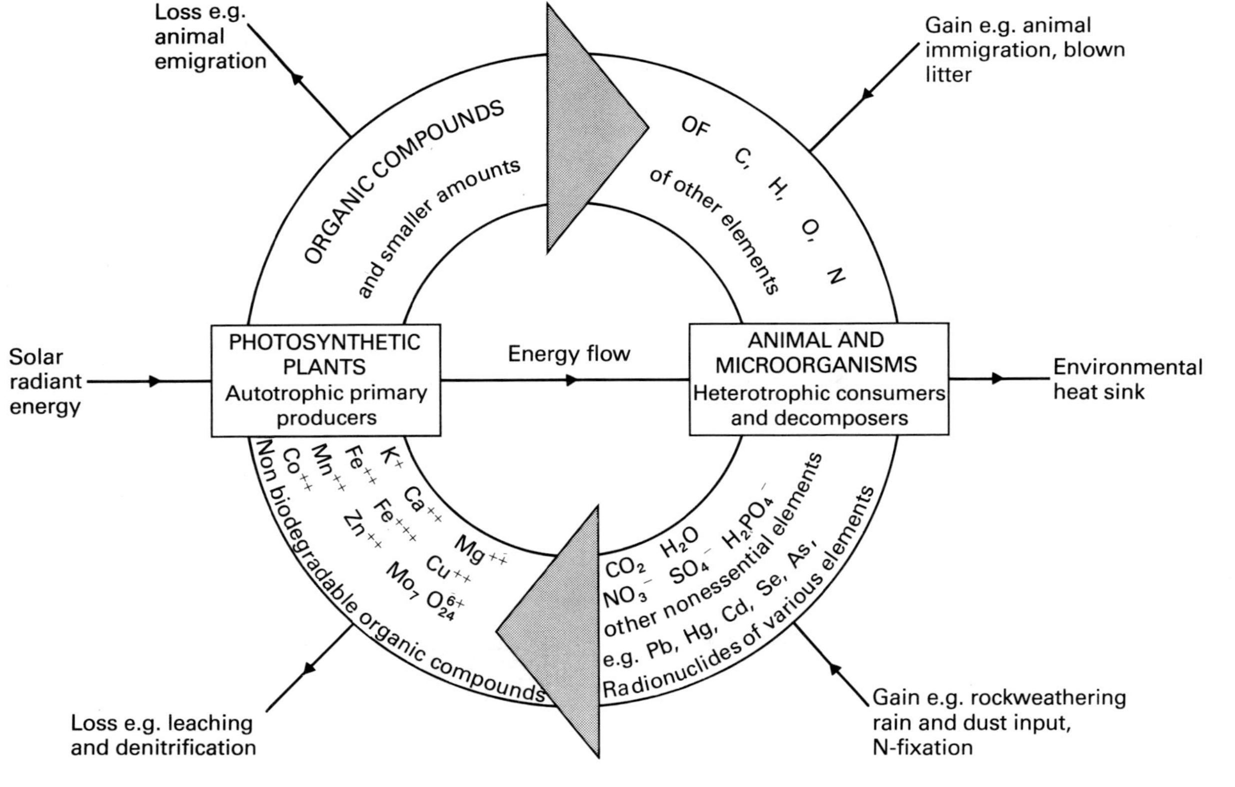

Loss e.g. animal emigration
Gain e.g. animal immigration, blown litter
ORGANIC COMPOUNDS OF C, H, O, N
and smaller amounts of other elements
Solar radiant energy
PHOTOSYNTHETIC PLANTS
Autotrophic primary producers
Energy flow
ANIMAL AND MICROORGANISMS
Heterotrophic consumers and decomposers
Environmental heat sink
K+ Ca++ Mg++
Fe++ Fe+++ Cu++
Mn++ Zn++ Mo7 O24 6+
Co++
Non biodegradable organic compounds
CO2 H2O H2PO4−
NO3− SO4−
other nonessential elements
e.g. Pb, Hg, Cd, Se, As,
Radionuclides of various elements
Loss e.g. leaching and denitrification
Gain e.g. rockweathering rain and dust input, N-fixation

the living and dead organic material of the ecosystem. These cycles may be classified into two types:

i. local cycles of elements which are not gaseous or have no gaseous compounds (Fig. 4–2a);
ii. global cycles of these elements which can circulate in the atmosphere in gaseous form (Fig. 4–2b).

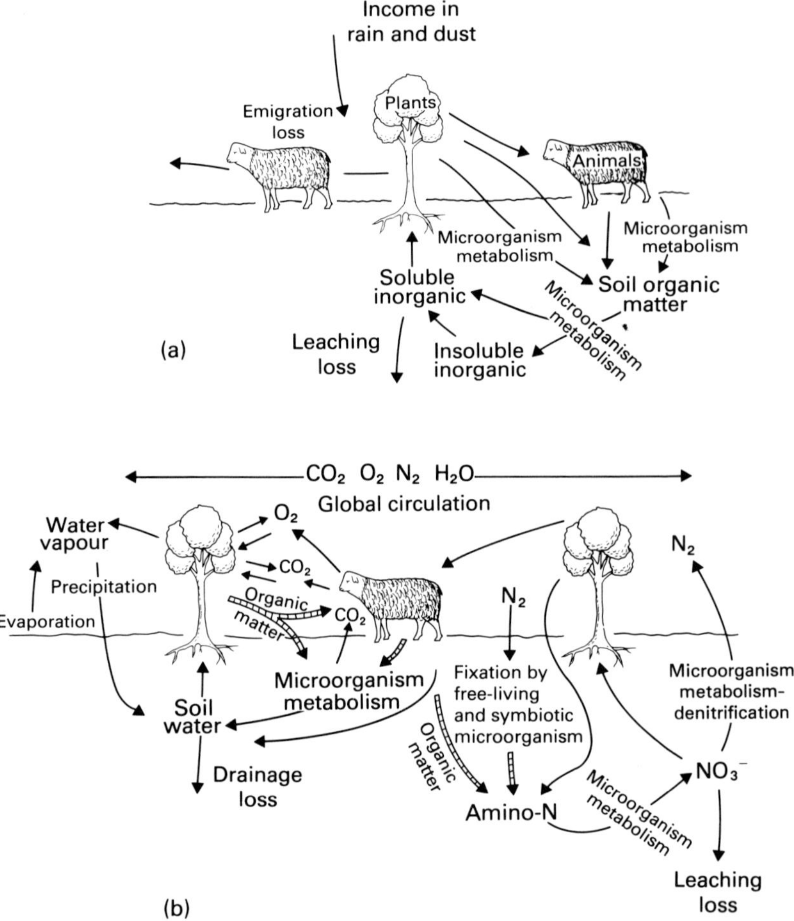

Fig. 4–2 Nutrient cycle types. (a) Local cycle with no gaseous components. Leaching loss is total, the materials removed finally reaching ocean or lake basins as sediments. Return to circulation only takes place over geological time periods. (b) Cycles of elements which are gaseous or have gaseous compounds. These are, perhaps not fortuitously, the elements which are required in largest amounts by living organisms to form the main structural components of organic molecules.

Local cycles include most soil-borne essential elements except nitrogen and also include those non-essential elements which are concentrated from soil solution by plants. Sulphur behaves generally as a localized element but, under anaerobic soil conditions, can be converted to gaseous hydrogen sulphide and enter the global cycle. The gaseous cycles are those of the structural elements which may circulate as oxygen, water or carbon dioxide, and of nitrogen which exists in the molecular form as an enormous atmospheric reservoir.

Elements with a global circulation are present at similar concentrations in most ecosystems, the composition of earth's atmosphere being, geographically, rather constant. An element such as nitrogen may still be limiting due to the slowness of a conversion process such as biological nitrogen fixation. By contrast, the local cycle elements are often in short supply and, if they are lost from the system by animal emigration or leaching, cannot be replaced except by slow release during the further weathering of geological parent materials. Such elements, when leached into water courses are lost to lake or ocean basins whence they may only be returned by geological processes. For this reason these cycles are alternatively known as *sedimentary cycles*.

4·3 Cycle limitation

The global cycles of hydrogen are largely in that part of the water cycle which does not involve living organisms. The biological cycles of oxygen and the part of the hydrogen cycle which involves passage of water through organisms, are dependent on earth's vegetation cover and its photosynthetic energy fixation.

The atmospheric concentrations of carbon dioxide (0.035%) and oxygen (21%) are, respectively, too low and too high for maximum photosynthesis. In the case of CO_2 this is simply due to a limited supply at high light intensities but, in the case of oxygen, the explanation is more complex. The normal 21% of O_2 in the atmosphere permits light-stimulated *photorespiration* to proceed unimpeded causing a reduction in apparent photosynthetic C-fixation. Further discussion appears in section 7.1.

Though nitrogen is available in enormous quantity from the atmospheric reservoir, the supply of nitrate to plants in most ecosystems is limited either by the rate of nitrogen fixation or by inadequate microbial mineralization of soil organic nitrogen.

The metal cationic nutrients such as potassium, calcium, magnesium and iron are present in soil solution at rather low concentrations. The reservoir which replenishes supplies is the exchangeable metal cation complement of the cation exchange complex which is in a transient-state exchange equilibrium with the surrounding solution. Losses caused by leaching and plant uptake or gains from litter decomposition are buffered by further ionization from, or recombination with, the exchange complex.

The anionic nutrients are not protected in this way as soils have little anion exchanging activity. For this reason it is fortunate that nitrogen, phosphorus and sulphur form organic compounds which are relatively resistant to microbiological breakdown so that the major reserve in many soils is in the water-insoluble organic form. For example nitrate usually forms less than 2% of soil nitrogen (BEAR, 1964) and, for this reason, measurements of microbial nitrification rates give a better estimate of nitrogen availability than the conventional Kjeldahl measurement of total-N.

The anionic nutrients are absorbed as nitrate, orthophosphate and sulphate ions and, consequently, the rate of microbial mineralization of the organic store may be limiting to plant uptake and to the rate of cycling. Some peat soils for example are very nitrogen-deficient and yet contain large stores of organic-N compounds. Phosphorus may also limit growth because, in alkaline soils, the solubility of calcium phosphate is low and in acid soils the solubility of iron and aluminium phosphates is very low.

4·4 Examples of nutrient cycles

Phosphorus is a good example of a local, or sedimentary cycle (Fig. 4–3). Phosphorus is usually present at very low concentration in rock minerals but the activity of plants and microorganisms during the soil-forming process results in fairly insoluble organic and inorganic phosphorus compounds becoming concentrated near the soil surface. Despite this, many ecosystems are phosphorus-limited because of the low solubility of its inorganic salts (section 4.3) or the resistance of its organic compounds to microbial breakdown. For the same reasons it is not rapidly lost by leaching.

Most other elements which have no gaseous cycle-component behave similarly, biological accumulation leading to greater concentrations of geologically uncommon elements in soils than in parent rocks. Such mechanisms also operate on non-essential elements, for example, heavy metals such as lead are more concentrated in the organic-rich surface layers of many soils. These effects are often exaggerated under nutrient-deficient conditions and may present a hazard when artificially created radionuclides enter ecosystems. Strontium 90 is, for example, accumulated by plants and cycled in a manner analogous to that of calcium. In upland, lime-deficient ecosystems it is strongly concentrated by plants whence it passes to the bones and milk of grazing animals.

The relative size of the organic, inorganic and living material pools of the various elements depends on such things as the relative

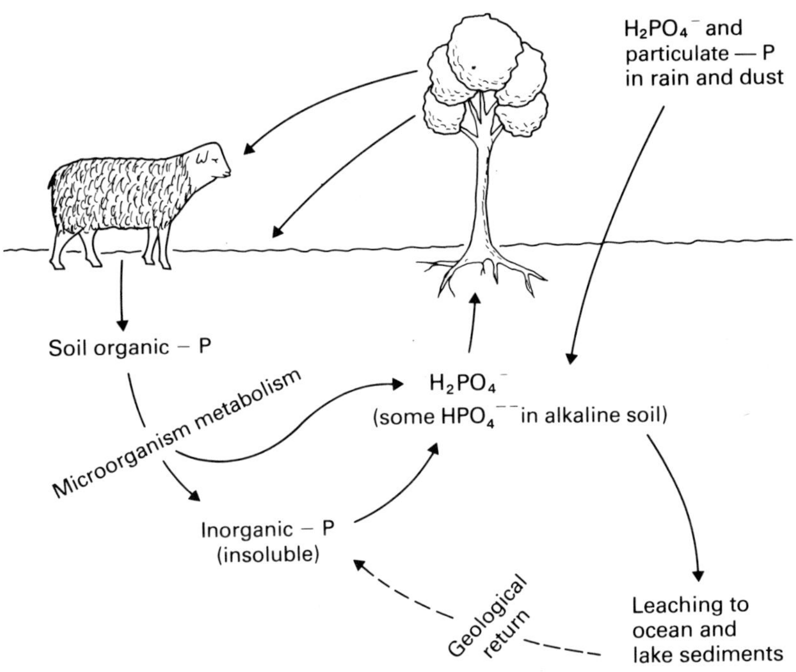

Fig. 4–3 The local cycle of phosphorus.

biodegradability of the organic compounds, the solubility of inorganic compounds and the degree of binding of different ion species to exchange sites. The predominance of monovalent ions in sea-water and divalent ions in soil solution and freshwater is a reflection of their relative mobilities in the biosphere.

The gaseous cycle of carbon (Fig. 4–4) is the central event of life on earth, the photosynthetic fixation of energy in carbon chain molecules being the root of all other life processes on the planet. The cycle is stable, showing a strong biological and chemical homeostasis. Carbon dioxide concentration is limiting to photosynthesis in the biosphere so that any increase in atmospheric content is compensated by increased photosynthetic uptake. The long-term homeostatic mechanism is, however, solubility in ocean water and the buffering effect of the carbonate-bicarbonate equilibrium.

Pure water at 15°C dissolves rather more than its own volume of carbon dioxide. This gas slowly hydrates to form carbonic acid which is then capable of almost instantaneous ionization to hydrogen ions, bicarbonate and carbonate anions, establishing equilibria which depend

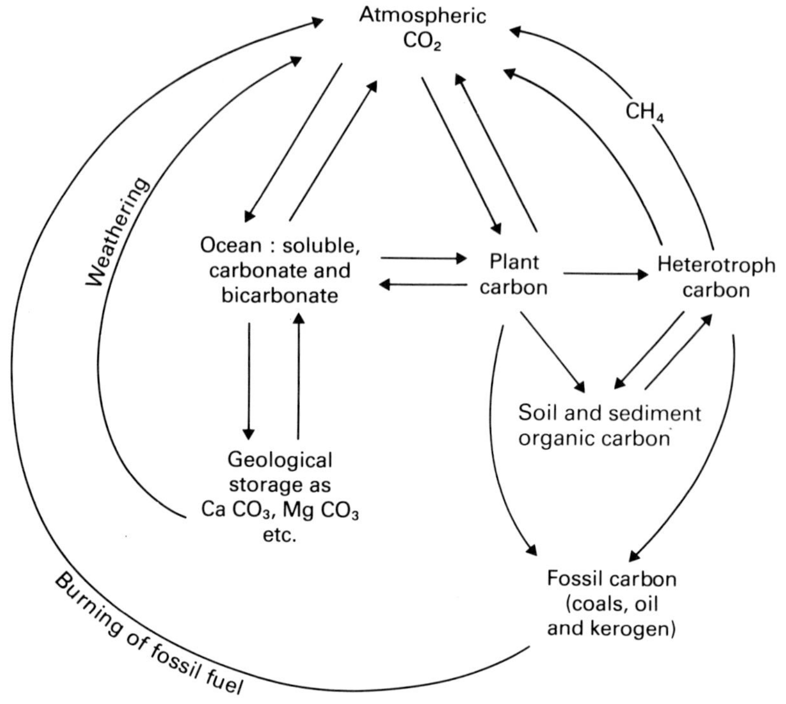

Fig. 4–4 The global carbon cycle.

on the partial pressure of carbon dioxide in the gaseous atmosphere above:

$$H_2O + CO_2 \leftrightharpoons H_2CO_3 \leftrightharpoons HCO_3^- + H^+ \leftrightharpoons CO_3^{--} + 2H^+$$

In the high salt environment of the ocean, calcium and magnesium carbonates may be close to saturation and precipitation, or biogenesis of calcareous shells, may occur. The concentration of carbon dioxide in the atmosphere is, in part, governed by these processes. Increasing concentration is counteracted by solution of carbonates:

$$CaCO_3 + CO_2 + H_2O \leftrightharpoons Ca(HCO_3)_2$$

Conversely, a reduction in carbon dioxide concentration, for example by photosynthetic withdrawal, is met by conversion of bicarbonate to carbonate. The total oceanic reservoir of dissolved inorganic carbon is more than 50 times that of the atmosphere but the equilibration time is slow because of limited transfer rates at the ocean surface and slow mixing of great depths of thermally stratified ocean water.

The gaseous cycle of nitrogen is more complex as it includes a number of different soil microbiological processes (Fig. 4–5). It is strongly

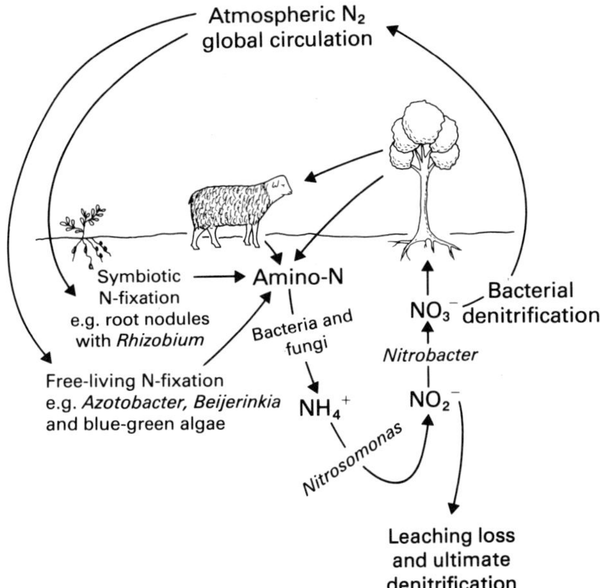

Fig. 4–5 The global nitrogen cycle. The enormous reserve of atmospheric N_2 provides an unlimited supply in all ecosystems but failure of microbial N-fixation, excessive denitrification or slow mineralization of organic N often cause limitation of plant growth.

stabilized by the large reservoir of atmospheric nitrogen, the ubiquity of nitrogen-fixing organisms and the efficiency of the denitrification process which avoids losses to sedimentary storage.

The cycle of sulphur is biologically interesting being of hybrid type (Fig. 4–6). The main cycle is sedimentary but there is a limited gaseous component due to bacterial production of hydrogen sulphide in anaerobic waterlogged soils, muds and marine sediments. Sulphur dioxide is also gaseous and has now been introduced to the sulphur cycle in considerable quantity by the burning of fossil fuel (section 4.5).

4·5 Cycles, pollution and eutrophication

The biogeochemical cycles vary in stability according to differences in capacity and rate functions. Local concentrations of human population have always influenced elemental cycles but the changes following the Industrial Revolution have extended this influence to the global scale.

The cycles of the biological structural elements, carbon, hydrogen and oxygen are all parts of global gaseous systems. It might be expected that carbon would be very sensitive to disturbance as the atmospheric

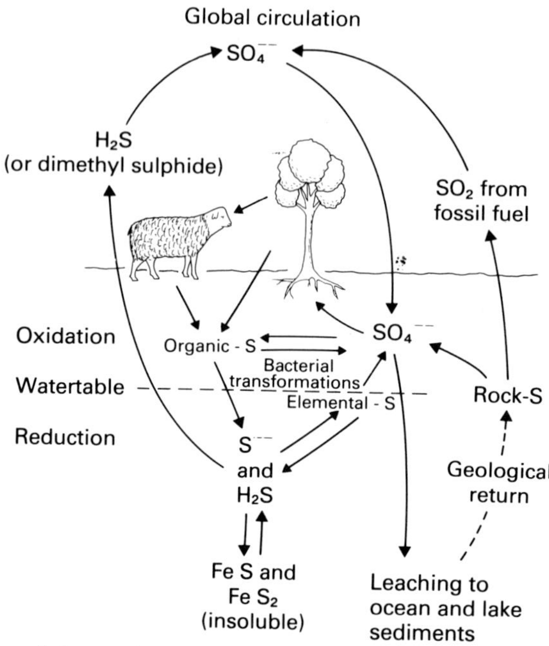

Fig. 4–6 The sulphur cycle. Because hydrogen sulphide is a gas and is formed in appreciable quantities in waterlogged soils and sediments, the cycle of sulphur is hybrid, behaving both as a local element within ecosystems but also having a global circulation component. It differs from the major gaseous cycles only in the proportion of the biologically available sulphur which enters the atmosphere.

concentration is so low (*c.* 330 vpm) but the homeostatic mechanisms described above have compensated for the release of very large quantities of fossil carbon since the Industrial Revolution. The atmospheric content has not increased by more than about 30–50 ppm during this century. It is unlikely that further release will directly influence the cycle but carbon dioxide in the atmosphere contributes to the 'greenhouse effect' which limits long-wave radiant loss from the earth's surface (section 2.6). It has been suggested that even a small increase in the atmospheric carbon dioxide content could interfere with the thermal and water balance of the atmosphere and similar fears have been expressed concerning the injection of large amounts of water vapour into the upper atmosphere from aircraft exhausts.

These potential disturbances are related to the structure and homeostatic function of the atmospheric system rather than to the cyclic nature of the processes involved. Similarly the interference with ozone in the upper atmosphere by nitrogen oxides in aircraft exhausts, and the interaction of nitrogen oxides, ozone and hydrocarbons to form peroxyacyl nitrates (PAN) in photochemical smog do not involve the direct disruption of cyclic systems.

The emission of sulphur dioxide into the atmosphere and the return of fossil carbon to the atmosphere-ocean system by the burning of fossil fuels, often considered to be 'air pollution', may also be interpreted as a rejuvenation process. Materials which were biologically immobilized in the geological past are being returned to active cycling by the agency of man. In the case of sulphur, the redispersion is on such a scale that the global input in rainfall is now much greater than the pre-industrial level of about 6 kg ha^{-1} y^{-1} (BEAR, 1964). Industrial production of sulphur dioxide does, however, cause serious local problems both by its toxicity to plants and by its damaging effect on human and animal respiratory tracts.

Phytotoxic materials like sulphur dioxide may interfere with nutrient cycles by blocking the energy-fixing photosynthetic process. Other toxins inhibit a whole range of metabolic processes and may either limit primary production or secondary microbial decomposition processes so reducing cycling rates. A different type of cycle disruption is caused by the entry of an unusually large quantity of a single element. The appearance of nitrate in run-off water from over-fertilized land or the introduction of phosphorus compounds from domestic and industrial detergents in sewage effluents are cases in point.

Accidental supplementation of major elemental cycles is named *eutrophication* and has the short-term consequence of disturbing population balance. In freshwater algal ecosystems a species which is normally limited by nitrogen, phosphorus or both may 'bloom' after eutrophication, seriously disturbing plant species balance, many less vigorous species being excluded by competition for light.

The long term consequence of eutrophication is that the organic matter, produced so rapidly during the bloom, sinks to the bottom and may exceed the oxidative capacity of the system. The bottom sediments may become exceedingly anaerobic causing extinction of many animal and plant species and, in a closed system like a lake, producing irreversible changes of trophic structure. A chain of events beginning with the supplementation of a single element thus culminates in the modification or near collapse of all elemental cycles. Further discussion may be found in MELLANBY (1972) E. P. ODUM (1971).

5 Limits to Growth

5·1 Limiting factors

The concept of limitation is encountered repeatedly in biological studies such as plant physiology and population growth. The idea first appeared in the mid-nineteenth century as Liebig's law of the minimum but BLACKMAN (1905), in a classic paper, *Optima and Limiting Factors*, extended the concept to encompass limitation by both maxima and minima. His 'law of limiting factors' states that: 'When a process is conditioned as to its rapidity by a number of separate factors, the rate of the process is limited by the pace of the "slowest" factor.'

The limiting-factor approach has been useful as an analytical tool because many chemical and physical factors may be controlled separately under experimental conditions. The pitfall is interaction: response to a single factor may be modified by variation in another and all conclusions from experiments must be carefully cross-checked by field observation of plant distribution and behaviour.

A further complication of the concept is that it is only applicable to steady-state conditions whereas most ecosystems are constantly changing in response to environmental factor variation, the life cycles of organisms and successional development. It is for this reason that the problem is currently being approached through systems analysis which permits time–course modelling of the consequences of simultaneous variation in many environmental factors (VAN DYNE, 1969).

Despite the move towards holistic ecosystem analysis, limiting-factor experimentation still has much to offer in establishing physiological–ecological profiles of individual species which furnish explanations for their field distribution and permit ecosystem modelling at the level of individual species response.

5·2 Limits: the physical environment

Terrestrial plants inhabit a physically demanding environment in which their aerial organs are subject to wide fluctuations of radiant energy flux, water deficit, temperature and air movement. By contrast their roots are embedded in a stabilizing mineral matrix which limits movement of gases, water and dissolved solutes, buffers rapid temperature change and so provides a more constant environment than the atmosphere above.

5.2.1 *Water deficit*

The combined effects of high radiant energy income, dry air and high temperature cause rapid evaporation. Because the plant leaf is an efficient gas-exchanging structure (section 2.2) it loses water vapour very easily unless the stomata are closed. This homeostatic mechanism is needed because the soil and plant-conducting pathway cannot keep pace with water demand during periods of high transpiration even if the roots are in moist soil.

Despite stomatal homeostasis, plant tissues show considerable diurnal variations of water deficit suggesting that either the soil or the plant-conducting pathway is limiting supply. As the soil dries after rain, the *soil water potential* falls so that the plant has to meet its water requirements against a gradually increasing energy gradient. (For discussion of water potential terminology see SUTCLIFFE, 1968.) At the same time the emptying of soil capillary pores reduces the cross-sectional area of the water transport pathway in the soil, increasing resistance to water movement (Fig. 5–1). The gradual increase in leaf water deficit is the plant response to the increasing energy gradient and decreasing water mobility as the soil dries. The diurnal oscillation represents the inability of the absorptive and conductive system to keep pace with demand during the day. At night the plant tissues re-equilibrate with the soil water potential.

The technique of *growth analysis* (EVANS, 1972) is particularly suited to showing the effect of water deficit on plants as it divides growth into two components: one reflects instantaneous changes in photosynthetic rate and the other defines the longer term morphological and anatomical impact of water deficit.

$$\text{relative growth rate} = \text{unit leaf rate} \times \text{leaf area ratio}$$

$$\frac{dW}{dt} \cdot \frac{1}{W} = \frac{dW}{dt} \cdot \frac{1}{A} \times \frac{A}{W}$$

where: W=plant weight; A=photosynthetic area and t=time. Water stress causes a reduction in photosynthesis attributable to stomatal closure, increasing mesophyll wall resistance to gas exchange or biochemical effects. This is manifested as a lowered unit leaf rate. Longer term effects appear as reductions of the leaf area ratio. This is a measure of the 'efficiency' with which plant tissue is deployed as a photosynthetic surface. Water deficiency reduces cell turgor and in an expanding leaf this will prevent cells from reaching their full potential size. A given weight of leaf tissue will thus produce a smaller photosynthetic area. This reduction of leaf area ratio by a short period of drought may cause a long term depression of relative growth rate from which the plant may never fully recover.

Water deficit has limiting effects other than reduction of growth.

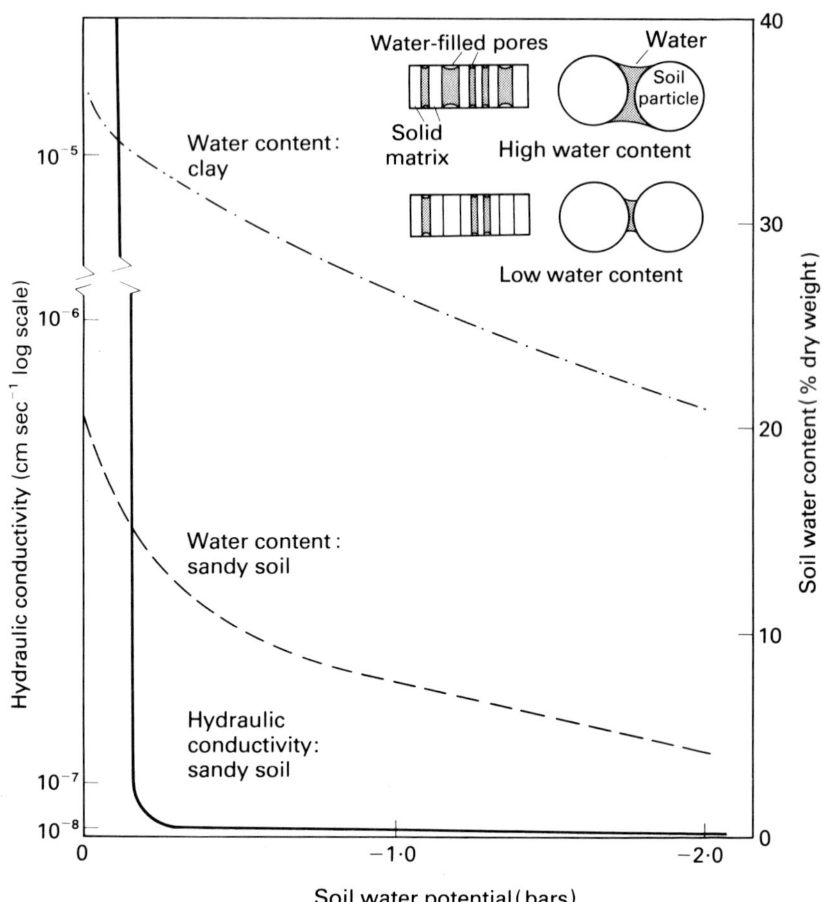

Fig. 5–1 The relationship between water conductivity in unsaturated soil, soil water content and soil water potential. Water conductivity and water content curves are shown for the same sandy soil and to show the influence of particle size, a second water content curve for a clay. The inset diagrams illustrate the reason for the very sharp inflection of the conductivity curve between 0.0 and −0.2 bar. Because of this very rapid rise in resistance water can move only very slowly through an unsaturated soil. After rain, for example, a wetting-front travels down a soil profile with very wet soil above and dry soil below.

Extreme drought may be lethal, for example the life cycles of many winter annuals may be terminated by drought in early summer. Water deficit may limit seed germination, impede essential element absorption from soil, reduce metabolite translocation, increase or decrease respiratory rate according to the degree of deficit and induce a wide range of biochemical, anatomical and morphological changes.

5.2.2 *Radiant energy*

The photosynthetic process depends on an adequate supply of radiant energy. Deficiency may arise through short day length, perhaps interacting with cloud cover or topography to reduce the integrated daily input to a value which is below the compensation point for some species. Natural selection will normally have resulted in plant distributions which avoid these problems. A commoner cause of light deficiency is shading by a canopy. Sunlight penetrates into a canopy and is exponentially reduced in intensity as it passes successive leaf layers. Leaves and plants in the lower strata of a deep canopy may be at sub-optimal light intensities but also show adaptation or specific selection to cope with these conditions. The differentiation of 'sun' and 'shade' plants in response to this situation is discussed in section 7.1.

A whole range of hormonally controlled functions such as dormancy, flowering, abcission, germination and many others are controlled by day length. The physiological basis of these responses, for example the phytochrome red : far-red mechanism, has been studied in some depth, but the deep ecological implications have been to a large extent ignored.

5.2.3 *Temperature*

Low temperature may be limiting because it depresses metabolic processes, photosynthesis for example being stopped in many species by temperatures near 0°C. Air temperature is not the sole factor to be considered. With variation of transpiration and radiation balance (section 2.2) plant temperature may be above or below ambient. Frost damage may thus occur even if air temperature remains a little above freezing and radiative heating may cause damage or metabolic distress in bright sunlight.

As with day length, temperature also has a strong control function, for example the breaking of dormancy in seeds and plants or the vernalization effect. Many of these are physiologically well defined but their ecological significance, though obvious, needs further study.

5·3 **Limits: chemical factors**

Chemical limitation is the consequence of deficiency or excess. Elemental deficiency is widespread while excessive amounts of some compounds or ion species are found in many ecosystems. Excess of water may induce toxic or otherwise adverse soil conditions and similarly, pH has a complex limiting relationship with plant growth through several different components of the soil chemical environment.

5.3.1 *Essential elements*

These may be subdivided into the structural elements, carbon, hydrogen and oxygen which are obtained from the atmosphere or from soil water and the essential mineral nutrient elements obtained from the soil. Of the structural elements, carbon is the only one likely directly to limit photosynthesis because its concentration in air is so low.

The mineral nutrient elements comprise six macro-nutrients – nitrogen, phosphorus, potassium, sulphur, calcium and magnesium – which are required in comparatively large amounts, and about seven micro-nutrients, including copper, zinc, boron, chlorine, molybdenum, manganese and iron, which are required in much smaller amounts.

Almost all natural ecosystems respond to the addition of N, P and K, either by growth increment or change of species composition suggesting that they are limited by these elements. The remaining macro-elements, particularly calcium are frequently the cause of limitation. For obvious reasons the micro-nutrients are less often limiting but deficiencies of all are recorded, at least for agricultural ecosystems. Deficiency symptoms such as chlorosis (Fe and Mg deficiency) and unusual leaf colourations are not so often seen in wild plants as in cultivars. The wild populations have undergone selection for some degree of tolerance whereas agricultural plants have been artificially selected for high fertilizer response and, consequently, requirement. Deficiency may arise for a number of reasons the most obvious of which is a limited supply in the soil parent material. A less obvious reason is failure of recycling from an insoluble organic form (sections 4.3 and 4.4) or limitation of uptake due to insolubility of an inorganic source.

Sandy soils tend to be deficient in the alumino-silicate clay minerals which are the main source of many plant nutrients and having a low cation exchange capacity are also prone to leaching loss of cations such as calcium, magnesium and potassium. Phosphorus may be immobilized in organic compounds which are resistant to microbial decomposition and may also be limited, at low pH, by the insolubility of iron and aluminium phosphates and, at high pH, of calcium phosphates. Other pH interactions may occur, for example low pH limits nitrogen-fixing bacteria and such soils tend to be nitrogen deficient.

5.3.2 *Toxicity*

Many of the micro-nutrients could, at higher concentrations, be re-christened 'toxic heavy metals'. Soils developed from metalliferous rocks or contaminated by mining and smelting operations may contain one or more of the 38, or so, metallic elements with a density greater than five. These are potentially toxic to most organisms, usually by damaging protein molecules and blocking enzyme systems. A good example of such toxicity is found in the 'serpentine barrens' developed on the chromium and nickel-rich metamorphic rock, serpentine. Only a few plant species are able to tolerate the toxicity and low calcium : magnesium ratio of these soils. Some species which are tolerant of a specific heavy metal have been used as geobotanical indicators in prospecting for ores.

Development of resistance to heavy metal toxicity by populations of plants which invade mining spoil or smelter waster may take place very rapidly by intense selection of tolerant individuals from seed populations. Some plant species are particularly fruitful of resistant individuals, the grasses *Agrostis tenuis, Festuca ovina* and *Anthoxanthum odoratum* often being the only plants to colonize spoils with high copper, lead, zinc and other metal concentrations (ANTONOVICS, BRADSHAW and TURNER, 1971).

Toxicity effects may be caused by low soil pH, for example the high exchangeable aluminium content of acid soils (section 3.3) has been shown to limit the growth of some calcicoles. Calcifuge species probably have a root chelation mechanism which protects them from excessive aluminium uptake but, because it may compete for iron, makes them sensitive to iron deficiency on high pH soils. Many Ericaceous plants such as *Rhododendron spp.* show strong lime-induced iron deficiency chlorosis on alkaline soils. A number of other toxicities related to waterlogging and environmental pollution are discussed elsewhere (sections 4.5 and 5.3.3).

5.3.3 Waterlogging

When a normal soil is flooded it passes through a series of changes induced by oxygen deficiency. The diffusion rate of oxygen in water is about 20 000 times slower than it is in air. Filling of the soil pore space with water prevents oxygen diffusion from the surface: root and microorganism respiration then removes oxygen from the system and a population of anaerobic microorganisms begins to develop.

During this process, and subsequently as the soil becomes progressively more anaerobic, its redox potential falls. At any given soil pH there are a number of critical redox values which mark the conversion of various oxidized ion species and compounds to a reduced condition. It is difficult to define these values as soils contain such a heterogeneous mixture of redox couples but PONNAMPERUMA (1972) suggests as a rough guide:

	redox potential corrected to pH 7 (millivolts)
oxygen undetectable	330
nitrate undetectable	220
manganese coming into solution ($Mn^{4+ \text{ and } 3+} \longrightarrow Mn^{2+}$)	200
iron coming into solution ($Fe^{3+} \rightarrow Fe^{2+}$)	120
sulphate undetectable ($SO_4^{2-} \rightarrow S^{2-}$)	−150

This time sequence of changes following waterlogging is also reproduced as a vertical profile from the surface of a waterlogged soil.

Accompanying these inorganic effects, a large number of organic compounds may appear in the soil, for example methane is often generated microbially in organic-rich paddy soils, sometimes forming a large proportion of the soil atmosphere. Ethene (ethylene), fatty acids, aldehydes, ketones and many other compounds may appear in anaerobic soils.

The complex system of reduced substances in waterlogged soils may seriously limit growth. Divalent iron, manganese, the sulphide ion and hydrogen sulphide are all very phytotoxic. Ethylene, if it is produced, has strong growth-controlling activity.

5.3.4 *Soil pH*

Natural soils range from pH 10, or above, if sodium carbonate is present, down to less than pH 3 when oxidation of sulphides has released free sulphuric acid. These extremes may be found in desert alkali soils and in drained marsh soils which contained copious amounts of ferrous sulphide. The more normal pH range is between about pH 3 in leached podsolic soils to about pH 8.4 in rendzinas with free calcium carbonate limestone fragments.

Within this range, soil pH *per se* is not normally limiting but has its influence through various nutritional mechanisms. Several of these have already been noted in sections 3.3 and 5.3.1. Deficiency of cationic nutrients, immobilization of phosphorus by iron and aluminium, failure of nitrogen fixation and toxic levels of soluble aluminium are common in acid soil. Alkaline soils often induce iron, manganese and phosphorus deficiencies as solubility effects.

5·4 Limits: biological effects

These include the effects of inter- and intra-specific competition, grazing, pollination, seed dispersal mechanisms and pathogenicity. The effect of predation on grazing intensity or the extremely complex interaction of pathogens, their vectors and vector-predators must have repercussions which are reflected throughout the whole ecosystem structure. Extremely painstaking detective work is often required to expose the causal mechanisms of most plant distributions.

6 Plant Growth and Distribution

6.1 Succession and adjustment

Each generation of plants, passing through a phase of sexual reproduction and seedling establishment, is exposed to the genetic filter of selection. Some individuals are better fitted than others to survive so that, in time, the ecosystem niches become filled with species or ecotypic populations which have least 'environmental friction'.

In the young ecosystem this process involves extensive changes in species composition accompanied by environmental modification. This *succession* may be interpreted as a gradual exploration of environmental possibilities by an increasingly diverse population of organisms. Once the system matures to a *climax* condition, energy fixation and dissipation appear to have reached equilibrium, the efficiency of nutrient cycling is maximized and changes with time are smaller and less directional, representing adjustment by each generation of plants to minor alterations in the environment.

6.2 Methods of physiological-ecological investigation

During the past half-century innumerable experiments have been undertaken to define the environmental requirements of various plant species. A large proportion of these have been agriculturally directed to the needs of the plant breeder, fertilizer technologist or irrigation expert. Many of the techniques of physiological ecology have been adopted from this type of work and Table 1 outlines the range of correlative and experimental work which may be undertaken.

Subjective observations of field distribution often lay the foundation for a subsequent experimental approach, for example the habitat notes in a flora contain such phrases as 'on wet mud' or 'usually on calcareous soil'. The conversion of such subjective information to a quantitative basis demands a carefully designed sampling programme in which the study area is sampled by randomly distributed quadrats. For each quadrat the presence/absence, cover or density of species X is recorded and the requisite environmental factor(s) measured. Figure 6–1 shows an example of this approach in which the occurrence of two *Galium* spp. has been related to soil pH in a leached Carboniferous Limestone grassland. Note that the mode of the *G. saxatile* distribution is at a lower pH than that of the samples containing neither species while that of the *G. sterneri (G. pumilum)* distribution is above it.

The relationship may be further explored by regressing the plant cover

Table 1 Correlation and experimentation as an aid to the interpretation of plant distribution.

	Field	*Laboratory*
Observation and correlation	Quantitative studies of plant distribution may be related to habitat types or to measured environmental factors. The data may be presented visually, for example, as distribution maps or the species abundance may be regressed against the intensity of environmental factors	Plants may be collected and returned to the laboratory for measurement of weight, size, morphological and anatomical characteristics, chemical and biochemical composition. These are all phenotypic measurements in which it is not possible to separate the consequences of plant genetic constitution from the formative effects of the environment
Experimental investigation of plant growth Weight- or size-increase, morphological change, physiological function, chemical or biochemical composition may be measured	1 Modification of natural habitat, e.g. by addition of nutrients, shading, irrigation, etc. 2*a* Single species grown with single or multifactorial environmental variation e.g. by addition of nutrients, shading, irrigation, etc. 2*b* Several species grown with single or multifactorial environmental variation 2*c* As 2*b* but including competitive interaction	3 Semi-controlled environments, e.g. glasshouse Single species or several species *a* Vary a single environmental factor *b* Multifactorial environmental variation 4 Full environmental control, e.g. plant growth cabinets Species and environmental combinations as in 3 5 As 4 but including competitive interaction

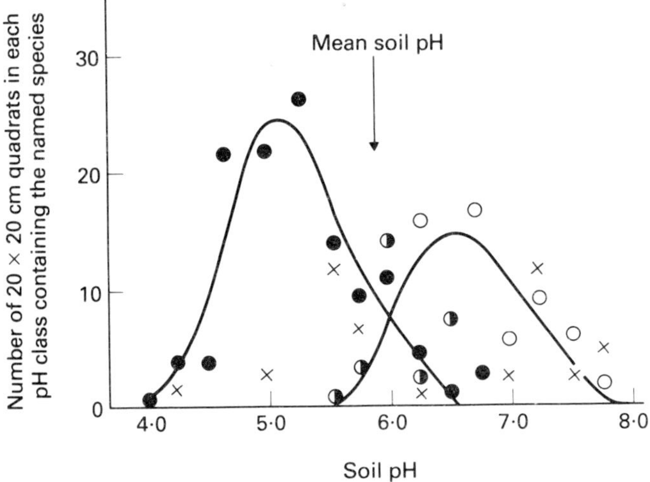

Fig. 6–1 Soil pH in quadrats containing *Galium saxatile* ●, *G. sternei* ○ both ◑ or neither **X** species. Grassland on shallow glacial drift overlying Carboniferous Limestone (Malham, Yorks).

data against soil pH as shown in Fig. 6–2. This strongly reinforces the impression that *G. sterneri* tends to 'prefer' the more alkaline soils but it provides no evidence of causal mechanism. It is not even legitimate to say that variation of pH causes the difference in distribution: it could be argued with equal force that the plants had modified the soil pH. A further problem in interpreting this particular correlation is that soil pH is, itself, so strongly correlated with many other environmental parameters (section 5.3.4).

A correlation never provides proof of causality, though circumstantial evidence may be very strong, but it does lay the foundation for subsequent experimental work. Some simple experiments with these two *Galium* spp. were described by TANSLEY long ago (1917), showing that the growth of each species was seriously inhibited by some factor in the soil from which it was normally excluded but no mechanism was suggested (see section 1.1).

The recent adoption of sophisticated multifactorial techniques for the analysis of vegetation community structure has introduced a new and valuable tool for the screening of the plant: environmental factor interaction as a precursor of experimental investigation. Ordinations are numerical techniques which permit the comparison of multifactorial variability between samples in such a way that it can be plotted within a geometrical reference frame bounded by two, three or more axes. The distance between the plotted sample points is then a measure of their similarity (small spacing) or dissimilarity (large spacing). In the case of

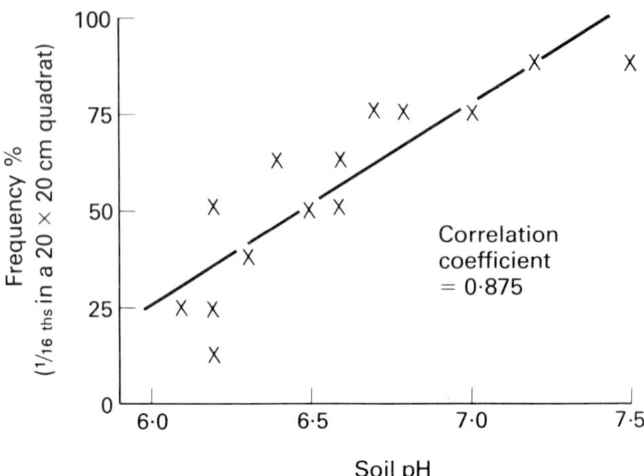

Fig. 6–2 The relationship between soil pH and the occurrence of *Galium sterneri*. The line fitted to the points is the calculated linear regression. (This and other statistical techniques are described in PARKER, R. E., 1973, *Introductory Statistics for Biology*, Studies in Biology no. 43, Edward Arnold, London.)

stands of vegetation the assessment of similarity may be made upon the species composition of the sample stands or upon environmental variables measured in the stands. A two or three-dimensional graph plot is easy to visualize but the *hypervolume* bounded by more than three axes can only be depicted by a set of separate comparison graphs. Ordination thus permits the separation of vegetation stands into a spatial framework which is a reflection of the variability in species-composition between each stand (Fig. 6–3). The distribution of a single species or the values of environmental variables may be projected onto the stand ordination giving a subjective impression of the ecological meaning of the ordination axes which may be further explored by regressing the intensity of each environmental factor against the position of the stand along a particular axis. KERSHAW (1973) provides a useful introduction to these concepts.

Establishment of hypotheses based on correlative data requires the measurement of many environmental variables. This is a topic which now has an enormous literature and cannot be reviewed here. Table 2, however, outlines some of the major techniques and gives reference to sources.

Once a hypothesis has been established, experimental testing is required. The majority of such work has been undertaken in glasshouse, or more critically controlled, environmental conditions. Few ecologically relevant field experiments have been undertaken with the exception of a wide range of trials with pasture species and varieties. Work with agricultural cultivars must be treated with reservation as many have been

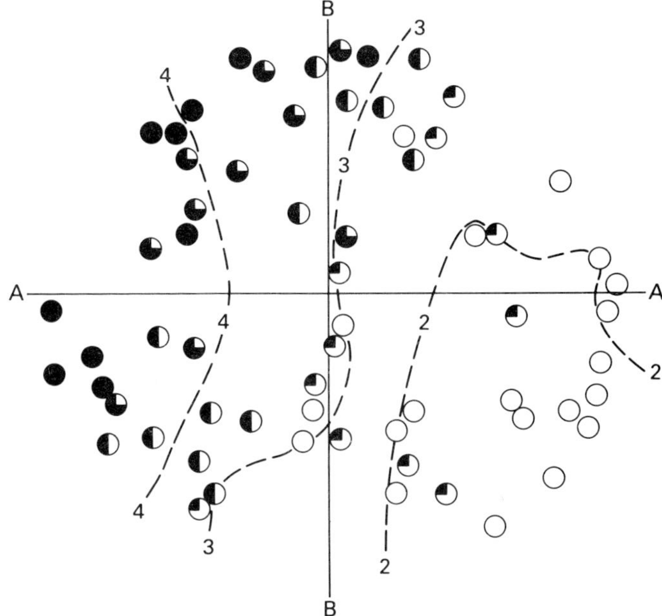

Fig. 6–3 A two axis ordination of cliff-top vegetation in Wales (after GOLDSMITH, 1973). The circles represent the position of individual vegetation stands in the ordination framework. The shading in the circles is frequency of *Armeria maritima* (thrift) expressed in quartiles. The 'contours' were plotted from the projection of soil conductivity into the stand ordination. Expressed in quartiles, the highest value of conductivity represents the greatest soil salinity and correlates with the highest frequency of the halophilous *A. maritima*. This is just one of the many species which could be projected onto the stand ordination in this way. (See also Fig. 7–3.)

selected for high nutrient response and other characteristics aimed at increasing yield. Many of them are not competitive in natural ecosystems and may have anomalous responses compared with wild varieties.

Comparative experiments are ecologically most valuable (GRIME, 1965) as they are ideally suited to the detection of specific susceptibilities which may explain distribution patterns. A simple comparative field experiment was described by RORISON (1960) who grew three plant species on three different soil types during an investigation of calcicole : calcifuge behaviour. Two calcicoles, *Scabiosa columbaria* (small scabious) and *Asperula cynanchica* (squinancy wort) germinated and survived well in a calcareous sand and in a chalk soil. In an acid sand they both germinated poorly and ultimately became extinct. The calcifuge, *Galium saxatile* (heath bedstraw) produced the opposite response, establishing a surviving population on the acid soil but becoming extinct on the others.

Table 2 Some commonly used methods of environmental measurement and source references. 1 – ALLEN, 1974; 2 – BLACK, 1965; 3 – CHAPMAN, 1976; 4 – CHAPMAN and PRATT, 1961; 5 – MONTEITH, 1972; 6 – SESTAK, CATSKY and JARVIS, 1971; 7 – WADSWORTH, 1968.
S – spot determination; R – recording; I – integrating. n-d – non-destructive; d – destructive; i – intermediate.

	Methods of environmental measurement	*Reference*
(*a*) *Radiant energy*	Solar short-wave input (0.3–3.0 μm): glass-domed thermopile radiometer. S. R. I. n-d	3 5 6 7
	All-wave, including thermal infra-red (3.0–100 μm): unshielded-ventilated or polythene-domed thermopile radiometer. S. R. I. n-d	3 5 6
	Net radiation (all-wave). Upward and downward facing all-wavelength radiometers electrically connected in opposition to each other. S. R. I. n-d	3 5 6
	Radiometers are usually calibrated to read energy flux density (e.g. W m^{-2}). The short-wave instruments may be filtered to separate infra-red from photosynthetically active radiation (PHAR)	
(*b*) *'Light'*	A wide range of electrical methods have been used to measure radiant energy in the visible part of the spectrum ('light'). They include: photoemissive and photomultiplier tubes (inert gas-filled or evacuated), photovoltaic cells (e.g. selenium barrier layer, silicon and germanium photodiodes), photoconductive cells (lead or cadmium sulphide and indium antimonide). S. R. I. n-d	3 5 6
	The use of such cells for biological purposes poses problems of spectral selectivity and careful choice of optical filtration to match response, e.g. to PHAR. Quantum detectors such as these may be used with instruments giving a response proportional to wavelength and thus reading quanta per unit wavelength. With the addition of a monochromator suitable detectors may be used for spectral analysis (spectroradiometry). Light detectors calibrated in photometric units (e.g. lux.)	
	Photochemical methods: polymerization of anthracene or uranyloxalate, exposure of photographic film or paper, diazo paper. I. n-d	6
(*c*) *Wind*	Cup or vane anemometers. S. R. I. n-d	5
	Manometric anemometers. S. n-d	
	Hot wire, heated thermocouple or thermistor anemometers. S. R. I. n-d	3 5
(*d*) *Temperature*	Mercury in glass thermometers. S. n-d	
	Thermocouples, thermistors, resistance thermometers. S. R. I. n-d	3 5 6 7

	Methods of environmental measurement	*Reference*
	Infra-red radiation thermometers and imaging devices for contactless measurement. S. R. n-d	6
(*e*) *Humidity*	Wet and dry bulb thermometers (psychrometer). S. n-d	
	Wet and dry bulb thermocouple, thermistor or resistance thermometers. S. R. I	3 5 7
	Conductance or capacitance hygrometer. S. R. n-d	5
(*f*) *Carbon dioxide*	Infra-red gas analysis. S. R. I. n-d	6
(*g*) *Precipitation*	Rain gauges. I or may be R. n-d	3 7
(*h*) *Soil chemical properties*	(i) Total elemental content. S. d	
	Na_2CO_3 fusion or HF solution of dry sample followed by colorimetry, emission or absorption spectrophotometry, selective ion or other electro-chemical analysis. N by Kjeldahl digestion	1 2 3 4
	Arc spectrography	1 2
	X-ray spectrography	1 2
	Electron probe and analytical electron microscopy have also been used	
	(ii) Exchangeable cations and cation exchange capacity. S. d	
	Displace exchangeable cations with, e.g., NH_4^+ Measure elements as in (h) (i). For exchange capacity measure NH_3 after distillation from soil – titration or selective ion electrode	1 2 3 4
	(iii) 'Available' nutrient elements. S. d	
	Various extractants have been used, generally because they have shown good correlation with crop response when a deficiency is indicated Water for NO_3^- and SO_4^{--}; various extractants including ion exchange resins for $H_2PO_4^-$. Measure elements as in (h) (i)	1 2 3 4
	(iv) Exchangeable hydrogen. S. d	
	Displace with a suitable cation at controlled pH and measure electrometrically or titrate	1 2 3 4
	(v) Soil pH (acidity). S. d	
	Electrometrically with glass-calomel toughened electrode in soil-water slurry or *in situ* in field Alternatively with indicators	1 2 3 4
	(vi) Organic carbon (organic matter). S. d	
	Wet oxidation with chromic acid and back titration or by ignition and measurement of CO_2 evolution	1 2 3 4
(*i*) *Soil physical properties*	(i) Water content. S. d	
	Weigh a sample, dry at 105°C and reweigh	2 3
	(ii) Water potential	

	Methods of environmental measurement	Reference
	Porous pot tensiometer with pressure gauge (to −1 bar only). S. R. n-d	2
	Porous electrical resistance blocks. S. R. n-d	2
	Neutron scatter or radiation absorption. S. R. n-d. d	2
	Thermocouple psychrometry (measurement of vapour pressure in equilibrium with soil). S. R. I	
	Laboratory calibration of soils by determination of water potential v. water content with suction plate and pressure membrane apparatus. S. d	2
(iii)	Particle size distribution (texture). S. d	
	Separation by sieving and sedimentation with gravimetric or hydrometric measurement	2
(iv)	Aggregate analysis (structure). S. d.	
	Wet sieving methods	2
(v)	Pore space. S. d	
	Capillary drainage or gas pressure/volume change	2
(vi)	Bulk density. S. d	
	Dry weight of a standard volume sample	2
(vii)	Penetrability. S. n-d	
	Hand penetrometer (e.g. for trampling compaction)	2

Although valuable, experiments such as this lack environmental control and are too multifactorial to permit more than general conclusions. The next step in such an investigation is a better controlled experiment to elucidate mechanisms. The work of PIGGOT and TAYLOR (1964) with *Urtica dioica* (stinging nettle) and *Mercurialis perennis* (dog's mercury) is a good example of such an experimental sequence. They observed that the two species often occurred separately as limestone woodland floor species but rarely together in mixed stands. Preliminary experiments suggested that the distribution might be caused by variation in soil phosphorus for which *U. dioica* has an unusually high demand. The results of a pot-culture experiment, undertaken to test this hypothesis are presented in Table 3.

In the soil which normally carries *M. perennis, U. dioica* shows a strong growth limitation unless phosphorus is added. Addition of nitrogen has no effect unless the phosphorus deficiency is first relieved. *M. perennis* shows no such differential response to phosphorus. The conclusion was finally checked in field germination and establishment trials of *U. dioica* in cleared plots which had carried *M. perennis*. Establishment failed without the addition of phosphorus.

Table 3 The growth of *Urtica dioica* and *Mercurialis perennis* on a soil from a *M. perennis* association either untreated or with nitrogen and phosphorus additions

		% maximum dry weight yield			
	Days of growth	Control	+ N	+ P	+ NP
U. dioica	33	5	4	74	100
M. perennis	88	69	93	96	100

Once experimental work has established a single factor explanation it becomes possible to investigate the physiological or biochemical mechanisms of differential response. Such experimentation often has the advantage of speed and permits the screening of large numbers of species. A good example is CRAWFORD and TYLER's study of respiratory adaptation to waterlogging (1969). After four days of experimental flooding they measured the 2-hydroxybutanedioic (malic) acid content of the roots of eleven plant species. The five tolerant species showed high levels of malate compared with much lower levels in the six intolerant species. The authors suggested that malate accumulates as a relatively non-toxic end product of anaerobic respiration which can be further metabolized when aerobic conditions return. The accumulation of malate results from the favouring of a lateral loop in the usual anaerobic pathway to ethanol from phosphoenol pyruvate (Fig. 6–4).

A pitfall of comparative experimentation is the genetic variability of different populations of the same species. Agricultural plant breeders have realized for many years that there is a wide pool of genetic variability which can be drawn on by imposing extreme selective pressures. It has now been shown for many wild species that a large number of factors may elicit population differentiation. Excellent examples are found in the literature of heavy metal tolerance (ANTONOVICS, BRADSHAW and TURNER, 1971). Races of such species as *Agrostis tenuis* (common bent grass) and *Festuca ovina* (sheep's fescue) which occur on mining and smelting waste have been naturally selected for their tolerance of the particular heavy metal(s) in the substrate.

Variations in the physical environment may be investigated by suitably controlled experimentation. A classic population study which also illustrates the use of such techniques is found in MOONEY and BILLINGS' (1961) investigation of the circum-polar arctic-alpine *Oxyria digyna* (mountain sorrel). Pot grown replicates from different arctic and lower latitude alpine habitats were exposed to a range of thermoperiod and temperature. As an example of the findings there was a clinal increase in photoperiod requirement for flowering from the southern to the

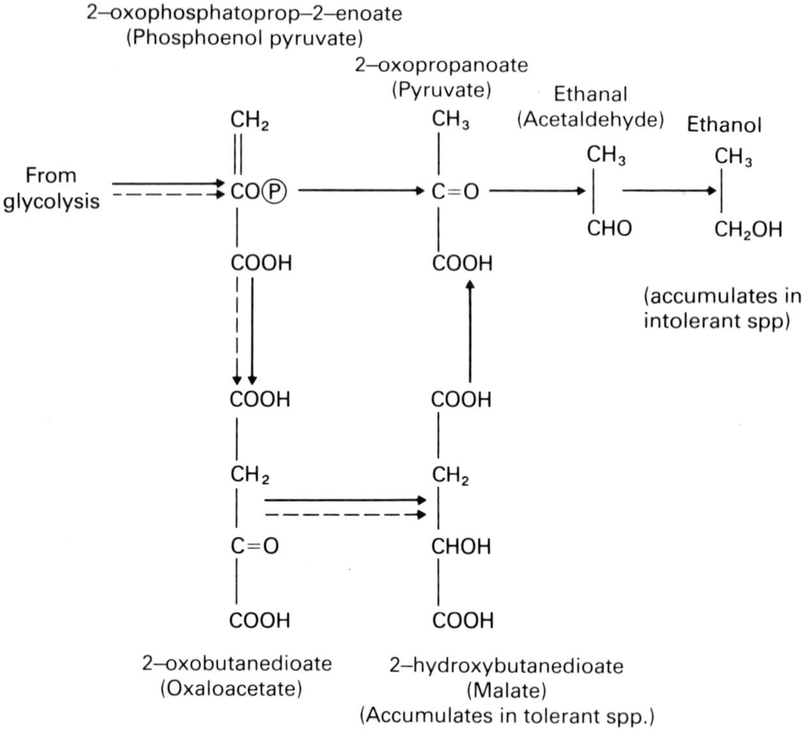

2–oxophosphatoprop–2–enoate
(Phosphoenol pyruvate)

2–oxopropanoate
(Pyruvate)

Ethanal
(Acetaldehyde)

Ethanol

From glycolysis

$CH_2 = C(O\text{Ⓟ}) - COOH \rightarrow CH_3 - C{=}O - COOH \rightarrow CH_3 - CHO \rightarrow CH_3 - CH_2OH$

(accumulates in intolerant spp)

$COOH - CH_2 - C{=}O - COOH \rightarrow COOH - CH_2 - CHOH - COOH$

2–oxobutanedioate
(Oxaloacetate)

2–hydroxybutanedioate
(Malate)
(Accumulates in tolerant spp.)

Fig. 6–4 Metabolic pathway of anaerobic respiration in waterlogging tolerant (– – –) and intolerant plants (——). (After CRAWFORD and TYLER, 1969.)

northern populations. From this and many other population differences the authors concluded that the wide range of the species was largely attributable to 'differences in metabolic potential among its component populations'.

7 Plants in the Ecosystem

7.1 Responses to limits

The previous chapter outlined some approaches to physiological–ecological investigation but gave no comprehensive treatment of the relationship between plant ecology and environmental conditions. Table 4 attempts to remedy this omission with some examples in this field.

No single response, selected by environmental pressure, can be considered in isolation. Many of the plant characteristics and environmental factors are interactive and many responses must be genetically linked together.

The complexity of such relationships is well illustrated by the interaction of biochemical photosynthetic strategy and water use efficiency in plants from different climatic regimes. Most temperate zone plants assimilate carbon through the Calvin-Benson cycle in which phosphoglycerate, a three-carbon compound, is the first product. A few grasses including *Saccharum officinale* (sugar cane) and *Zea mays* (maize), various members of the Amaranthaceae, Chenopodiaceae and some other families have evolved an alternative four-carbon strategy in which the first product is 2-oxobutanedioic (oxaloacetic) acid. These plants also lack light-induced *photorespiration* and are able to reduce substomatal CO_2 concentrations to 0–5 vpm compared with the 50–100 vpm of C–3 plants. A third strategy, Crassulacean acid metabolism (CAM), is found in some arid-zone plants which fix CO_2 at night by formation of malic acid from 2-oxopropanoic (pyruvic) acid via oxaloacetic acid. During the day the stomata close and the CO_2 is released to the Calvin-Benson cycle by decarboxylation of the malic acid.

CAM is a water conservative strategy which permits daytime stomatal closure and contrasts with the C–3 system in which any daytime closure will reduce CO_2 fixation. At any given intercellular CO_2 concentration, C–4 plants are able to photosynthesize faster than C–3 plants and consequently tolerate some degree of stomatal closure which reduces water loss without seriously limiting photosynthesis. Many C–4 plants also have higher temperature optima than most C–3 plants. These three strategies, quite predictably show an ecological correlation, CAM being essentially a desert plant mechanism while C–4 plants are more frequent in low latitude semi-arid ecosystems with C–3 predominating in wet equatorial areas and high latitudes.

Table 4 How plants cope with limits

Limiting factor	*Characteristic tending to be genotypically selected or phenotypically developed*
PHYSICAL	
1 *Water deficit*	*(a)* Deep rooting or rapid root growth into moist soil *(b)* Increased root : shoot ratio *(c)* Xeromorphic modifications including reduced surface to volume ratio, leaf rolling, hair coverings and water storage tissue *(d)* True drought resistance with cytoplasmic tolerance of desiccation (poikilohydry; only common in lower plants) *(e)* Combination of physiological characteristics giving maximum photosynthesis and minimum transpiration at a given leaf water potential, e.g. CAM (2 (c) iii below) *(f)* Evasion by dormancy (plant or seed) or part shedding *(g)* Large reserve of seed in soil and staggered germination *(h)* Adult plants widely spaced, e.g. by allelopathy
2 *Radiant energy*	*(a)* Maximization of photosynthesis at any level of radiant input by optimization of canopy architecture (i) Leaf area index (LAI = leaf area/ground area) selected to give maximum photosynthesis under prevailing conditions, e.g. complex stratification of forest ecosystems (also involving physiological and other differences – see (b) ii and (c) iii below) (ii) Optimization of 'display' to give minimum self-shading, e.g. by near vertical leaves or leaf-mosaics. Over-illumination avoided by edgewise orientation of leaves *(b)* Optimization of photosynthetic response by anatomical and morphological characteristics (i) Limiting mesophyll thickness avoiding cellular self-shading in low-light habitats (ii) Maximization of leaf area ratio (leaf area/plant weight) and specific leaf area (leaf area/leaf weight) under low light conditions *(c)* Optimization of photosynthetic response by physiological means (i) Minimization of respiration/photosynthesis in shade leaves and plants (ii) Hatch-Slack photosynthetic pathway and absence of photorespiration permitting high light saturation values, high photosynthetic yield in high light regimes and water conservative photosynthesis with partially closed stomata (iii) Crassulacean acid metabolism (CAM) permitting photosynthesis in very dry conditions (iv) Light saturation at low intensity in shade plants. (v) Photoperiod and light exposure effects adjusting flowering, germination, etc., to prevailing climate

Limiting factor	*Characteristic tending to be genotypically selected or phenotypically developed*
3 *Temperature*	*(a)* Selection for respiration/photosynthesis balance which maintains highest productivity despite greater sensitivity of respiration than photosynthesis to temperature change (tropical forest consumes a greater proportion of its gross production by respiration than does temperate forest) *(b)* Resistance to freezing damage either by 'antifreeze' mechanisms or molecular resistance to cytoplasmic freezing *(c)* Resistance to high temperature damage, probably at the molecular level *(d)* Mechanisms limiting radiative heating *(e)* Thermoperiod and exposure effects optimizing vernalization, dormancy breaking, etc., in relation to environmental timing
4 *Wind*	*(a)* Morphological creation of self-shelter by wind-pruning, streamlining in wind, prostrate life-form, flexibility and good root anchorage *(b)* Anatomical selection for or induction of xeromorphic characteristics *(c)* Active regeneration from basal parts
5 *Fire*	*(a)* Thick bark *(b)* Buried (geophyte) or surface (hemicryptophyte) perennating organs and active regeneration from basal parts *(c)* Large reserve of seed in soil *(d)* Seed dormancy broken by fire *(e)* Rapid exploitation of bare soil by seed or vegetative spread
CHEMICAL	
6 *Essential elements*	*(a)* Individual deficiencies: characteristics favouring tolerance. (i) Survival and growth with limited uptake (ii) More efficient or rapid uptake by root systems than that of competitors (iii) Exploitation of different soil layers (iv) Development of N fixing symbiosis (v) Rhizosphere microflora which aids absorption *(b)* Tolerance of adverse ionic ratios, e.g. low Ca^{++}/Mg^{++} on Serpentine or Dolomite, high Na^{+}/other metals in saline soil *(c)* Tolerance of high concentration of individual ions giving competitive advantage by creating a refuge, e.g. halophytes (Na^{+}) and calcifuges (Al^{+++})
7 *Waterlogging*	*(a)* Good internal aeration (aerenchyma) permitting sufficient oxygen diffusion from aerial parts to support root respiration and also oxidation of toxic, reducing materials in the rhizosphere

Limiting factor	*Characteristic tending to be genotypically selected or phenotypically developed*
	(b) Tolerance of toxic Fe^{++}, Mn^{++}, S^{--}, and reducing organic compounds *(c)* Tolerance of low O_2 and high CO_2 in soil *(d)* Development of unusual anaerobic respiratory pathways *(e)* Reduction in amount of respiring tissue in plant body, e.g. by porous aerenchymatous structure *(f)* Limitation of transpiration by xeromorphic leaf characteristics – reduces rate of entry of toxic materials *(g)* Tolerance of low NO_3^- supply due to denitrification and slow N-fixation. N absorbed as NH_4^+
8 *Soil acidity (low pH)*	*(a)* Tolerance of induced or associated deficiences such as P, Ca, Mg, K, etc., and failure of N Fixation *(b)* Tolerance of associated Al^{+++} toxicity *(c)* Tolerance of assorted organic toxins produced under these conditions *(d)* Tolerance of secondary biotic effects such as litter accumulation and failure of seedling establishment
9 *Soil alkalinity* (high pH)	*(a)* Tolerance of induced Fe and P deficiency *(b)* Tolerance of high ratio of Ca to other metals
BIOTIC	
10 *Grazing*	*(a)* Buried (geophyte) or surface (hemicryptophyte) perennating organs *(b)* Prostrate growth habit; flowers on short or horizontal stalks. Often coupled with slow growth rate *(c)* Toleration of trampling damage to plant and soil *(d)* Rapid vegetative regrowth *(e)* Large reserve of seed in soil *(f)* Unpalatability – morphological or biochemical *(g)* Rapid exploitation of bare soil by seed or vegetative spread
11 *Pathogenicity (fungi, bacteria and viruses)*	*(a)* Structural resistance to penetration *(b)* Biochemical resistance (i) Genetic capacity for development of resistant races (ii) General resistance in the species *(c)* Morphological or anatomical 'containment' by abscission or 'sealing-off' *(d)* Tolerance of infection: many wild populations are pathogen saturated but remain alive and reproduce *(c)* Resistance to activity of vectors
12 *Pollination and seed dispersal*	*(a)* Adjustment to available local organisms by increased diversity of utilization *(b)* Adjustment of timing of life cycle *(c)* Competitive attraction of agency

7.2 Competition or interaction

When two organisms each require more than a half share of an environmental resource, they compete. One may be driven to extinction, both may survive on sub-optimal intake or one may take a larger proportion and establish a strongly dominant but stable relationship with the other.

Plants compete for a range of environmental factors of which the most important are radiant energy (light), carbon dioxide, nutrients, water and, to a lesser extent, space. A few active modifications of the environment by secretion of specific toxins (allelopathy) may be important in some circumstances. Competition is strongly multifactorial and may be governed by the interaction of many environmental factors.

The modifications and behaviour which make a plant an efficient competitor are generally those favoured by the selective pressures outlined in Table 4. Probably the central factor in competition is the need for light. It is unique in that it cannot be stored and must be instantaneously intercepted and used.

The efficiency of leaves at different light intensities is very variable both within and between species and is also prone to acclimation effects. The result is that a multispecies canopy is a most efficient utilizer of light and the intensity at the canopy base may be less than one per cent of that incident above and yet shade leaves at this level still maintain a positive balance of photosynthesis.

Because the interception of light is so important it has been suggested that most other limiting factors ultimately operate through competition for light. The limitation of *Urtica dioica* by soil phosphorus, described in section 6.2 is a good example. *U. dioica* is a good competitor while the phosphorus supply is adequate but, at lower levels it makes inadequate height growth and, below a critical value, will be shaded out by *Mercurialis perennis*.

Competition is so potent a force in the ecosystem that many adaptations favour avoidance rather than tolerance. The differentiation of aspect societies in deciduous woodland is a case in point. The early spring 'picture postcard' assemblage of *Anemone nemorosa* (wood anemone), *Endymion non-scriptus* (bluebell) and *Mercurialis perennis* (dog's mercury) produces leaf very early and achieves most of its annual photosynthesis before the tree canopy closes.

In competition for nutrients and water, root stratification plays a similar role to the separation in time described above. The floor vegetation of nutrient deficient deciduous woodlands, dominated by *Pteridium aquilinum* (bracken) with *Holcus mollis* (creeping soft grass) and *E. non-scriptus*, shows a marked root stratification, three different levels being tapped for nutrients and water.

A rather different avoidance mechanism is found in the process of refuging. HACKETT (1967) showed that *Deschampsia flexuosa* (wavy hair grass) has a rather slow growth rate and low phosphorus requirement and is tolerant of high soil aluminium levels. Grown alone it performs best in fertile soils of mid-pH but its natural habitats are acid heathland and acid woodland floors. Hackett suggested that its low growth rate makes it a poor competitor but the nutrient-demanding species which outcompete it in fertile soil cannot tolerate the phosphorus deficiency and aluminum toxicity of the soils which consequently serve as a refuge for it. Almost any specific tolerance of a well-defined environmental stress, such as metal toxicity, high salt concentration, nutrient deficiency, water deficit or exposure may be interpreted in this sense.

7·3 Niche differentiation

Consider a model system containing two species one of which is less efficient in extracting soil phosphorus. As its individuals die and return organic phosphorus to the soil so the more efficient species will succeed in accumulating more and more of the environmental phosphorus reserve and the inefficient species will utimately become extinct. In the real world it is inconceivable that two species should show exactly balanced requirements for environmental resources and yet common observation is that natural ecosystems are species diverse and usually, the most mature and stable systems show great diversity. What is the explanation for the collapse of diversity and its replacement by monospecific dominance in the model and the maintenance of diversity in the real ecosystem?

One mechanism by which two species may cohabit is selection for differential use of resources whereby the population growth of one species becomes limited by a level of a specific resource which is not limiting to the other species. Figure 7–1 illustrates the development of self density dependence because of nitrogen deficiency in species **A** before it has exhausted the whole of the soil phosphorus reserve and driven species **B** to extinction.

This type of relationship which HARPER (1967) has described as 'focusing the intensive battles within rather than between species' is one key to the avoidance of exclusive struggles which would establish monospecific systems. Each species has available a large number of permutations and combinations of environmental 'choices' within a complex space–time system. Provided that a single species uses just one choice which does not overlap with those of its competitors it will achieve density dependent population control before coming into exclusive competition with other species.

This utilization of environmental choice to permit coexistence has been termed *niche differentiation*. Each species has a 'physiological profile' which defines its maximum range of habitat occupancy but competition will impose a niche confinement so that its ecological tolerance range will

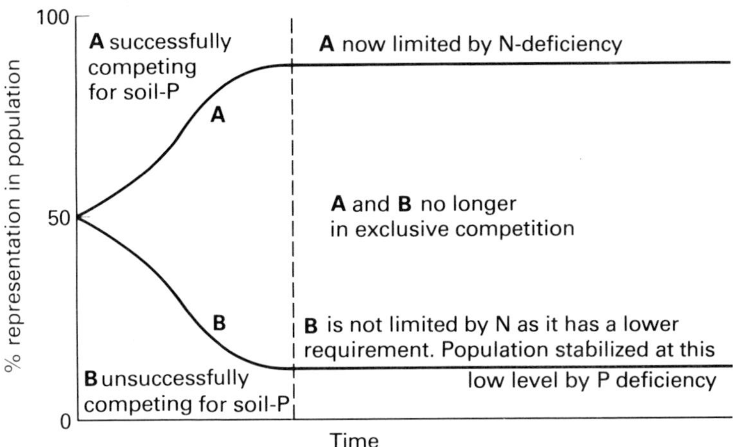

Fig. 7–1 The cohabitation of two plant species and its relationship to limitation by two nutrient elements. Species **A** is a better competitor for soil-P than species **B** but has a higher N requirement and so cannot entirely exclude species **B** by competitive stress.

represent only a part of the full physiological tolerance range. The case of refuging in *D. flexuosa* is a good example, competition in fact confining it to habitats which are not physiologically optimal.

Niche filling is a long term ecological process related to succession and to modification of the environment by successive seral communities. For example the halophyte association of British salt marshes has been stabilizing since the last glaciation but an open niche remained in the lower tidal mud which permitted the explosive spread of *Spartina anglica* (common cord grass) after about 1890. This species arose as a fertile amphiploid by the doubling of the chromosome number of *S.* × *townsendii*, a sterile hybrid of the British *S. maritima* and American *S. alterniflora* (HUBBARD, 1968). The only native competitor is *Salicornia stricta* agg. (glasswort) which is a small, rather weak annual quite unfitted to compete with the vegetatively aggressive *S. anglica*. By producing a new niche-filler, human activity has radically altered many British salt marshes.

7·4 Experimental investigation of competition

Habitat modification experiments such as those suggested in Table 1 often provide clues to the competitive status of wild ecosystems for example by delivering one or more species from limitation. WILLIS (1963) made factorial additions of NPK fertilizers to a dune grassland and showed that the niches of many small creeping and rosette perennials are preserved by nutrient limitation of potentially vigorous grasses such as *Festuca rubra* (red fescue) and *Poa pratensis* (meadowgrass). The algal 'blooming' of a eutrophicated lake is an accidental release from nutrient limitation in which the niches of many other species may be destroyed.

A most useful technique of paired species experimentation was introduced by DE WIT (1960) (cited by DONALD, 1963). Propagules or seed of two species may be planted together in varying ratios which maintain a constant density of individual plants, for example 50% : 50% or 75% : 25% in the planted mixture. Figure 7–2 shows two possible results of such an

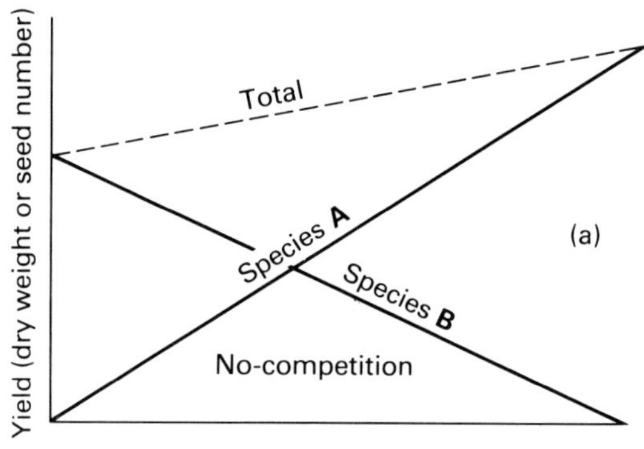

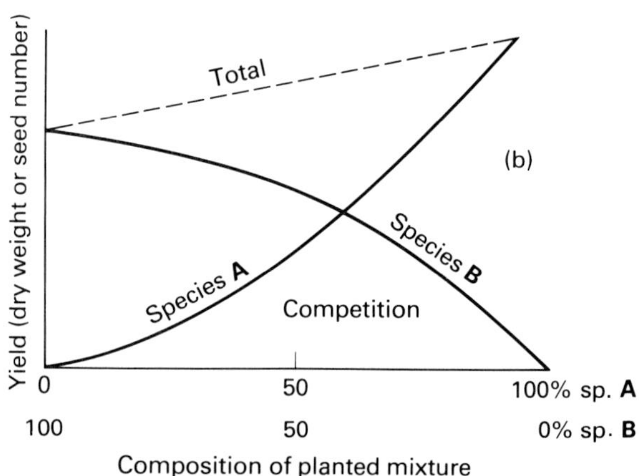

Fig. 7–2 Competition between two species sown or planted in different ratios to the same total density. (**a**) Different growth or seed production rates but no competition. (**b**) Growth or seed production rate as in (**a**) but competitive interaction.

experiment. Figure 7–2a is the situation in which relative growth rates or seed production differ but there has been no actual competition. In Fig. 7–2b the suppression of yield of species **A** is evidence that it is being outcompeted. This elegant technique is ideally suited to investigation of wild species relationships in different environmental conditions. Figure 7–3 shows the results of an experiment in which *Festuca rubra* (red fescue) and *Armeria maritima* (thrift) were competed with, and without sea-water irrigation. Of these two cliff plants, *A. maritima* is more halophilous and proved a better competitor in the sea-water treatment but its role was reversed in the control. This experiment was undertaken to confirm the

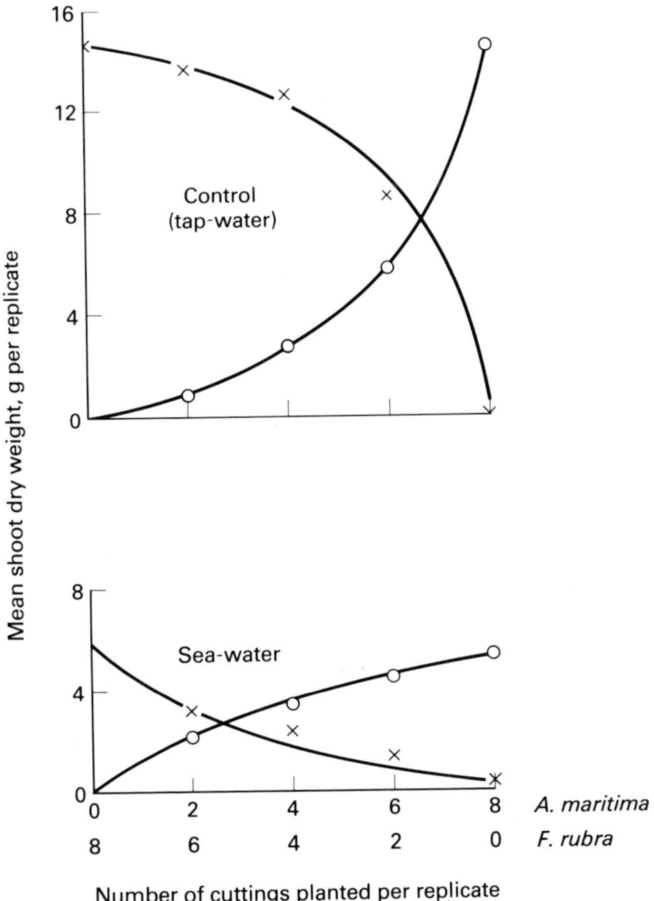

Fig. 7–3 Competition between *Armeria maritima* (**O**) and *Festuca rubra* (**X**) in a ratio experiment with either tap-water or sea-water irrigation. (After GOLDSMITH, 1973.)

interpretation of the ordination analysis of field data shown in Fig. 6–3. (GOLDSMITH, 1973).

This technique tells something of relative yield during a single season's growth but it is also possible to use the results to predict the long term consequences of competitive inhibition. Relative reproductive rates (RRR) may be calculated from ratio experiments:

$$\text{RRR} = \frac{\mathbf{A}_{\text{harvest}}/\mathbf{A}_{\text{sown}}}{\mathbf{B}_{\text{harvest}}/\mathbf{B}_{\text{sown}}}$$

If the community is stable and the seed ratios of the two species do not change from sowing to harvest then the value of RRR is unity and plotting the relationship of harvest mix to sown mix, as in Fig. 7–4a, produces a

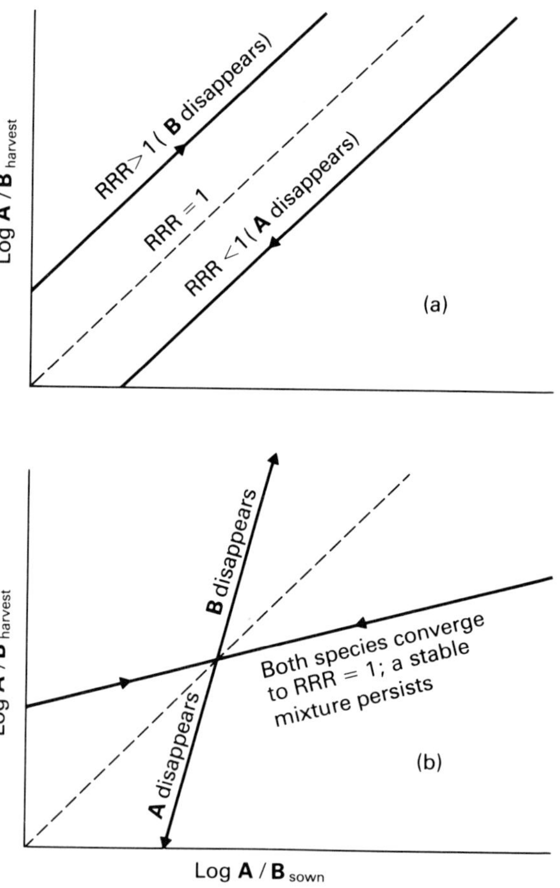

Fig. 7–4 Theoretical relationships between planted and harvested seed ratios of two species, **A** and **B**, grown in competition.

line with a slope of 45° passing through the origin. If one species produces more seed than the other, irrespective of sowing ratio, then the plotted line will be above or below but parallel to the 45° line. With the passage of seasons one species will disappear as shown by the trend arrows.

If seed production is affected by ratio dependent competition then the relationship departs from 45° (Fig. 7–4b). If it is greater than 45°, one species or the other will drift to extinction but if it is less, the mixture converges upon a stable RRR value which lies on the 45° line through the origin. If one species produces more seed than the other, irrespective of sowing ratio, then the plotted line will be above or below but parallel to the 45° line. With the passage of seasons one species will disappear as shown by the trend arrows.

It is most unlikely that any pair of species will have the 45° relationship as this demands exactly unchanged seed ratios from sowing to harvest but, of the four other possbilities, only one leads to stable cohabitation. This is the ratio-dependent situation in which, if one species is over represented in the sown mix, its proportion will fall by harvest time but if it is under represented then the converse will happen. This is the situation in which at a critical population density a species begins to inhibit its own growth, more than it inhibits its competitor. This is a condition for niche differentiation discussed in section 7.3 and, on an ecosystem scale, would lead to the adjustment of population levels so that the balance between reproduction and mortality would result in a constant species composition.

References

ALLEN, S. (1974). *Chemical Analysis of Ecological Materials*. Blackwell, Oxford.

ANTONOVICS, J., BRADSHAW, A. D. and TURNER, R. G. (1971). Heavy metal tolerance in plants. *Adv. ecol. Res.*, **7**, 1–85.

BEAR, F. E. (1964). *Chemistry of the Soil*. Reinhold, New York.

BLACK, C. A. (1965). *Methods of Soil Analysis*, Vols. 1 and 2. Amer. Soc. Agron., Wisconsin.

BLACKMAN, F. F. (1905). Optima and limiting factors. *Ann. Bot.*, **19**, 281–95.

BUNTING, B. T. (1965). *The Geography of the Soil*. Hutchinson, London.

CHAPMAN. H. D. and PRATT, P. F. (1961). *Methods of Analysis for Soils, Plants and Waters*. University of California, Riverside.

CHAPMAN, S. E. (1976). *Methods in Plant Ecology*. Blackwell, Oxford.

CLEMENTS, F. E. (1905). *Research Methods in Ecology*. Nebraska University Publishing Co., Lincoln, U.S.A.

COOPER, J. P. (1975). *Photosynthesis and Primary Production in Different Environments*. Cambridge University Press, Cambridge.

CRAWFORD, R. M. M. and TYLER, P. D. (1969). Organic acid metabolism in relation to flooding tolerance in roots. *J. Ecol.*, **57**, 237–46.

DIMBLEBY, G. (1967). *Plants and Archaeology*. John Baker. London.

DONALD, C. M. (1963). Competition amongst crop and pasture plants. *Adv. Agron.*, **15**, 1–117.

EVANS, G. C. (1972). *The Quantitative Analysis of Plant Growth*. Blackwell, Oxford.

GATES, D. M. (1962). *Energy Exchange in the Biosphere*. Harper and Row, New York.

GOLDSMITH, F. B. (1973). The vegetation of exposed cliffs at South Stack, Anglesey. I and II. *J. Ecol.*, **63**, 787–818 and 819–29.

GRIME, J. P. (1965). Comparative experiments as a key to the ecology of flowering plants. *Ecol.*, **46**, 513–15.

HACKETT, C. (1967). Ecological aspects of the nutrition of *Deschampsia flexuosa* (L). Trin. III. *J. Ecol.*, **55**, 831–40.

HARPER, J. L. (1967). A Darwinian approach to ecology. *J. Ecol.*, **55**, 247–70.

HUBBARD, C. E. (1968). *Grasses*, Penguin Books, Harmondsworth.

KERSHAW, K. A. (1973). *Quantitative and Dynamic Plant Ecology*, 2nd edn. Edward Arnold, London.

LIETH, H. and WHITTAKER, R. H. (1975). *Primary Productivity of the Biosphere*. Springer Verlag, Berlin.

LINDEMAN, R. L. (1942). The trophic dynamic aspect of ecology. *Ecol.*, **23**, 399–418.

McNEIL, M. (1964). Lateritic soils. *Sci. Amer.*, **211**, 5, 96–102.

MELLANBY, K. (1972). *The Biology of Pollution*. Studies in Biology no. 38. Edward Arnold, London.

MONTEITH, J. L. (1972). *Survey of Instruments for Micrometeorology* (I.B.P. Handbook 22). Blackwell, Oxford.

MONTEITH, J. L. (1973). *Principles of Environmental Physics*. Edward Arnold, London.

MOONEY, H. A. and BILLINGS, W. D. (1961). Comparative physiological ecology of arctic and alpine populations of *Oxyria Digyna*. *Ecol. Monogr.*, **39**, 1–29.

MOORE, P. D. and BELLAMY, D. J. (1973). *Peatlands*. Elek Science, London.

ODUM, E. P. (1971). *Fundamentals of Ecology*, 3rd edn. Saunders, Philadelphia.

ODUM, H. T. (1957). Trophic structure and productivity of Silver Springs, Florida. *Ecol. Monagr.* **27**, 55–112.

ODUM, H. T. (1971). *Environment, Power and Society*, Wiley Interscience, New York.

PAUL, E. A., CAMPBELL, C. A., RENNIE, D. A. and McCALLUM, K. J. (1964). Investigation of the dynamics of soil humus utilizing carbon dating techniques. *Trans. 8th. Int. Congr. Soil Sci.*, 201–8.

PENMAN, H. L. (1963). *Vegetation and Hydrology*. Comm. Bur. Soils Tech. Comm., 53.

PIGGOT, C. D. and TAYLOR, K. (1964). The distribution of some woodland herbs in relation to the supply of nitrogen and phosporus in the soil. *J. Ecol.*, **52** (supplement), 175–85.

PHILLIPSON, J. (1966). *Ecological Energetics*. Studies in Biology no. 1. Edward Arnold, London.

PONNAMPERUMA, F. N. (1972). The chemistry of submerged soils. *Adv. Agron.*, **24**, 29–96.

RORISON, I. H. (1960). Some experimental aspects of the calcicole-calcifuge problem. I. *J. Ecol.*, **48**, 585–99.

SESTAK, Z., CATSKY, J. and JARVIS, P. G. (1971). *Plant Photosynthetic Production: Manual of Methods*. Junk, The Hague.

SWANK, W. T. and DOUGLASS, J. E. (1974). Streamflow greatly reduced by converting deciduous hardwood stands to pine. *Science.*, **185**, 857–9.

SUTCLIFFE, J. (1968) *Plants and Water*. Studies in Biology no. 14. Edward Arnold, London.

TANSLEY, A. G. (1917). On competition between *Galium saxatile* L. (*G. hercynicum* Weig.) and *G. sylvestre* Poll. (*G. asperum* Schreb.) on different types of soil. *J. Ecol.*, **5**, 173–9.

TANSLEY, A. G. (1939). *The British Islands and their Vegetation*. Cambridge University Press, Cambridge.

VAN DYNE, G. M. (1969). *The Ecosystem Concept in Natural Resource Management*. Academic Press, N. York.

WADSWORTH, R. M. (1968). *The Measurement of Environmental Factors in Terrestrial Ecology*. Blackwell, Oxford.

WILLIS, A. J. (1963). Braunton Burrows: the effects on the vegetation of the addition of mineral nutrients to the dune soils. *J. Ecol.*, **51**, 353–74.

WIT, C. T. DE (1960). On Competition. *Versl. landbouwk. Onderz. Ned.*, **66**, 8.

Additional Reading

BENNET, D. P. and HUMPHRIES, D. A. (1974). *Introduction to Field Biology*. 2nd edn. Edward Arnold, London.

CLARKSON, D. T. (1974). *Ion Transport and Cell Structure in Plants*. McGraw-Hill, London.

COLLIER, B. D., COX, G. W., JOHNSON, A. W. and MILLER, P. C. (1974). *Dynamic Ecology*. Prentice Hall, London.

EPSTEIN, E. (1972). *Mineral Nutrition of Plants: Principles and Perspectives*. Wiley, London.

ETHERINGTON, J. R. (1975). *Environment and Plant Ecology*. Wiley, London.

GIMINGHAM, C. H. (1972). *Ecology of Heathlands*. Chapman and Hall, London.

KREBS, C. J. (1972). *Ecology*. Harper International, New York.

LARCHER, W. (1975). *Physiological Plant Ecology*. Springer Verlag, Berlin.

MUELLER-DOMBOIS, D. and ELLENBERG, H. (1974). *Aims and Methods of Vegetation Ecology*. Wiley, London.

RANWELL, D. S. (1972). *Ecology of Salt Marshes and Sand Dunes*. Chapman and Hall, London.

SLAYTER, R. O. (1967). *Plant-Water Relationships*. Academic Press. London.

STALFELT, M. G. (1972). *Plant Ecology: Plants, the Soil and Man*. Longman, London.

WATTS, D. (1971). *Principles of Biogeography*. McGraw-Hill, London.

WILLIS, A. J. (1973). *Introduction to Plant Ecology*. Allen and Unwin, London.